Electricity Restructuring

The Texas Story

L. Lynne Kiesling and
Andrew N. Kleit, Editors

The AEI Press

Publisher for the American Enterprise Institute

WASHINGTON, D.C.

To order call toll free 1-800-462-6420 or 1-717-794-3800.
For all other inquiries please contact the AEI Press, 1150 Seventeenth Street, N.W., Washington, D.C. 20036 or call 1-800-862-5801.

NRI National Research Initiative

This publication is a project of the National Research Initiative, a program of the American Enterprise Institute that is designed to support, publish, and disseminate research by university-based scholars and other independent researchers who are engaged in the exploration of important public policy issues.

Library of Congress Cataloging-in-Publication Data

Electricity restructuring : the Texas story / L. Lynne Kiesling and Andrew N. Kleit, editors.

p. cm.

Includes bibliographical references and index.

ISBN-13: 978-0-8447-4282-3

ISBN-10: 0-8447-4282-1

1. Electric utilities—Texas. 2. Electric power—Texas. I. Kiesling, Laura Lynne, 1965– II. Kleit, Andrew N.

HD9685.U6T437 2009

333.793'209764—dc22

2009020463

13 12 11 10 09 1 2 3 4 5 6 7

Contents

List of Illustrations

Figures

Tables

Preface

Brett A. Perlman

The movement to open retail electricity markets to competition in the United States has been largely written off by politicians, policymakers, pundits, and the press. Following the infamous meltdown of the California energy market in 2000–1 and the corporate collapse of Enron in 2001, these restructuring efforts came to an abrupt halt.

The political effects of the crisis reverberated across the country as electricity deregulation became politically electrified, a "third rail" issue so charged that public officials touched it at their own peril. The 2003 recall election of Gray Davis—in which the citizens of California took the virtually unprecedented step of removing a recently elected sitting governor, in part as a result of his handling of their state's electricity crisis—was only the first example of this political turmoil. In Maryland, the state legislature tried to "fire" its Public Service Commission when the commission approved a 72 percent retail rate hike following the end of price caps imposed by Maryland's 1999 electricity competition plan. Maryland Governor Robert Ehrlich suffered the same fate as Gray Davis, losing a 2006 reelection bid in which electricity issues were a major factor.

Sensing the potentially dangerous political consequences of competitive electricity markets, other states exercised damage control. The Public Utilities Commission (PUC) of Ohio, concerned with the dearth of competitive electricity suppliers and market activity, found in 2006 that market-based rates were not in the best interests of Ohio customers. It then implemented "rate stabilization plans" that stretched out rate increases from higher energy costs but reimposed regulated prices at levels that prevented competitors from entering the market. And it repealed electricity restructuring altogether in 2008.

Similarly, in 2007, the Virginia legislature repealed its electricity competition law and re-regulated its utilities, even though there had never been a meaningful effort to establish a deregulated market in Virginia.

Retail electricity competition has fared no better among pundits or the press. The libertarian Cato Institute, whose stated mission promotes "traditional American principles of limited government and free markets," published a piece by its own scholars that recommended the total abandonment of existing restructuring efforts; instead, the institute declared, "Those states that have already embraced restructuring [should] return to an updated version of the old, vertically integrated, regulated status quo."[1] Both the *New York Times* and the *Wall Street Journal* have printed stories stating that competition has failed to shrink electricity bills and showing that states are rolling back initiatives to restructure their markets. Regional and local newspapers have focused similarly on the failure of retail competition in the electricity industry.

Thus, the conventional wisdom appears to conclude that competition does not work in retail electricity markets.

Yet this orthodoxy ignores the Texas experience in electricity restructuring. The Texas electricity restructuring story is significant not only because of its successful track record, but also because of the size, importance, and characteristics of the state's electricity market. If Texas were its own country (as Texans would sometimes prefer), it would rank as the eleventh largest electricity consumer in the world, ahead of Australia, Mexico, Spain, and South Korea. Texas is also unique because its Public Utility Commission is the state's only utility regulator. The PUCT determines both retail and wholesale rules for the Electric Reliability Council of Texas (ERCOT) region, which covers 85 percent of the state and is home to over 5.6 million customers. These numbers make the scale and scope of the Texas market larger than most other restructuring programs around the globe. In this sense, the Texas regulatory experience more closely resembles electricity restructuring in other countries, such as the United Kingdom, Australia, and Chile, where electricity markets have flourished. Ultimately, it stands as a successful model of wholesale and retail electricity restructuring alongside these international examples.

Called "dereg" by statehouse lobbyists, the Texas electricity program was not, strictly speaking, "deregulation," but one of the most lengthy, complex, and comprehensive restructuring projects ever undertaken in any deregulated industry in the United States. The effort began in 1995 with the opening of the ERCOT wholesale market to competition, which allowed independent power producers to build new power plants without seeking regulatory approval. This new regulatory framework encouraged market entrants to build over 20,000 megawatts (MW) of new, efficient, combined-cycle gas generation capacity, which was seen as a cleaner and cheaper alternative to coal and nuclear power plants.

In 1999, the Texas legislature completed the job by passing the landmark law, Senate Bill 7, which extended restructuring to retail electricity markets. This bill created a market restructuring process in which the state's investor-owned, vertically integrated utilities were split into separate business units, and it established a five-year transitional period that began in 2002 with market opening and ended in 2007 with full deregulation of retail electricity prices, making Texas the only state with completely unregulated retail electricity prices. Unlike many other states, Texas successfully completed electricity restructuring in the ERCOT power region on schedule; and even though electricity deregulation has been controversial in Texas, there has not been, as in other states, any concerted movement to return to regulation since the retail market opened.

This is not to say that electricity restructuring in Texas has not caused heated policy debates. Residential prices skyrocketed following the retail market opening, and many blamed deregulation, although the timing of retail restructuring coincided with a period of increases in natural gas prices. A highly controversial proposal in 2006 by TXU, the state's largest generation company, to build eleven conventional coal-fired power plants without plans to mitigate greenhouse gases caused an international furor, which triggered the sale of the company in the largest private equity takeover in history and even attracted Hollywood star Robert Redford to finance an independent film called *Fighting Goliath: Texas Coal Wars* about the incident. Renewed electricity price spikes and the financial market meltdown in 2009 have forced several retail providers out of the market.

In spite of these controversies, the consensus for continuing electricity competition appears, by and large, to have held among state legislators, the governor, and the state utility commission. Moreover, Texas suffered none of the repercussions that occurred in other states when transitional price controls ended as planned in January 2007, in part because the PUCT set prices at market levels during the transitional period.

What we can learn from the Texas example is not limited to the political dimension, however. The Texas market has been successful based on many important objective indicators across a number of dimensions:

- *Retail market entry.* The Texas retail market has low entry barriers that have provided opportunities for unusually large numbers of competitive providers. According to the PUCT, as of 2008, there were over sixty active suppliers providing over 100 products in the market.
- *Customer switching.* Since Texas provided no price protection to industrial customers at market opening, virtually the entire industrial-customer load switched almost immediately to competitive offers. In addition, a significant number of residential customers chose competitive providers, which serve over 50 percent of the residential load, as of 2009. Over 80 percent of residential customers were taking a competitive product offering, even if provided by an incumbent supplier.
- *Power generation.* Power generation companies have built over 25,000 megawatts of new gas-fired generation, representing an investment of over $25 billion, since 1995. As of 2009, ERCOT has a project queue of over 100,000 megawatts—all built or to be built with private capital and without any risk to Texas customers.
- *Fuel diversification.* The market apparently is providing incentives to diversify Texas's power supply, with three large generation companies vying to build the first nuclear power plant in the United States in over twenty-five years, as well as the only nuclear power plants ever to be proposed anywhere in the world with private-risk capital and no ratepayer support.

- *Renewable power generation.* By 2009, Texas led the United States in wind energy development, with close to 9,000 megawatts of existing renewable resources and a transmission development program that could support up to a total of 18,000 megawatts of wind projects.

Yet the Texas electricity restructuring story has had minuses along with these pluses. One commonly heard criticism is that Texas retail electricity prices have become extremely volatile, since the state currently depends heavily on power plants supplied by natural gas. Indeed, retail electricity prices surged as natural gas prices reached all-time highs in 2006, following hurricanes Katrina and Rita, and in 2009, and they have remained persistently high as a result of the run-up in gas prices during the middle part of the decade. This is more an outgrowth of the state's reliance on natural gas as its primary fuel source for electric power generation, however, than of the underlying electricity market design, which simply reflects the prices created in the wholesale electricity market. Thus, the complaint is akin to blaming the umpire for the outcome of the game.

Some electricity policy purists have criticized Texas for failing to heed lessons learned elsewhere in its decision to adopt a zonal wholesale market design rather than the nodal design preferred by academics and implemented in the eastern U.S. energy markets. Although the zonal design traded pure economic efficiency for operational simplicity, Texas policymakers have since decided to make a transition to a nodal design by the end of 2010. Yet the zonal versus nodal debate seems to miss the broader picture: The ERCOT market has created a sustainable wholesale and retail market structure, an accomplishment beyond that of other U.S. electricity markets. Given its apparent success, the ERCOT experience may cause economists to rethink some basic wholesale electricity market design tenets. As one respected market pundit has quipped, the key question for energy economists is not whether the ERCOT market works in practice, but how it can possibly work in theory.

Moreover, economists must consider whether the regulatory, economic, or policy lessons learned from the Texas experience can be applied in other markets. Or whether—as some believe—Texas is an anomaly, because of its unique political environment, market, and regulatory structure.

This book represents the first systematic attempt to reflect on the Texas restructuring experience and to assess both its successes and shortcomings. *Electricity Restructuring: The Texas Story* tells how the state successfully implemented its electricity restructuring model at a time when most others were practicing damage control and rapidly pulling back from electricity restructuring, and it shows how public officials can successfully work with stakeholders, through both good and difficult times, to achieve an important policy objective. The book also demonstrates, in the most classic sense, the federalist principle that states can serve as laboratories for innovation if policymakers are not afraid to try new market innovations, even when leading economists suggest they will not work.

At the same time, the Texas electricity experience raises questions about some basic free-market principles: Why do companies and individuals not always make decisions that objectively appear to be in their own best interests? Why, for example, have not all residential customers switched to lower-cost competitive products that are exactly the same as higher-cost products? Why have small generators not taken advantage of clear opportunities to maximize their profits by participating in the Texas wholesale electricity market? In exploring cases in which individuals do not act as economists expect, the book provides fodder for challenging the conventional economic thinking about how energy markets are supposed to work. Indeed, the book offers a rich set of experiences for future study, not only of electricity economics, but also of energy policy, economic theory, and industrial organization.

In Texas, we like to say that electricity restructuring was "done right." *Electricity Restructuring: The Texas Story* shows how it was done.

Introduction

L. Lynne Kiesling and Andrew N. Kleit

Starting in the late 1960s, microeconomics had one policy prescription for achieving economic efficiency in capital-intensive infrastructure industries: deregulation. Deregulation had important successes to its credit in airlines, railroads, trucking, natural gas, and petroleum. So, when the cost of electricity began to rise in many states due to cost overruns on nuclear power plants, microeconomists came back to their favorite solution. After all, the regulated utility structure was based on the assumption that the relevant stages of production were natural monopolies. By the early 1990s, however, it was clear that the generation and marketing of electricity were not natural monopolies and that the cost-based justification for continued regulation of a vertically integrated industry was disappearing.

Following this realization, a number of states moved toward (but in most cases did not achieve) the deregulation of the generation and marketing of electric power. Because the steps were only partial and were suffused with political compromises, they met with, at most, mixed success. For most political actors, however, the end of electricity restructuring came with the California debacle of 2000–1. There, a poorly designed restructured system, which bore only a passing resemblance to a free market, melted down and imposed large costs on California ratepayers; the result was the early retirement of Governor Gray Davis. Since the California escapade, several states have moved backward with electricity restructuring, and no state has moved forward.

No state, that is, except Texas. In particular, in the Electric Reliability Council of Texas (ERCOT) area, the Public Utility Commission of Texas

(PUCT) has moved forward with a bold restructuring effort that has opened up both wholesale and retail markets for electricity. *Electricity Restructuring: The Texas Story* tells how Texas, alone among U.S. states, moved forward into a truly restructured and competitive electricity era. Unlike most other analyses of electricity restructuring, the volume combines academic analyses with firsthand accounts and retrospective surveys of restructuring from some of the actual architects of the restructuring design.

The overall success of electricity restructuring in Texas raises two important questions: Why was it successful? And how can that success inform restructuring processes in other states (and countries)? In large part, the contributions to this volume indicate that success in Texas was due to three factors: a competitive vision and the political leadership to carry it through; an institutional design whose transparent rules enabled decentralized coordination; and ongoing regulatory analysis of market outcomes, combined with the willingness to use that analysis to revise market rules and facilitate competition.

The economic vision that informed Texas's electricity restructuring was grounded in one simple, yet powerful, idea: Market processes and competition do a better job than political processes in harnessing private knowledge to reduce long-run costs, increase consumer choice, and encourage innovation. Market processes are positive-sum games because they generally make both consumers and producers better off than they would have been otherwise. This statement does require two caveats: No institutional change in the real world makes all consumers better off; and those producers who benefit from regulation—that is, incumbent utilities—have strong incentives to oppose restructuring and stymie its progress. It took strong political leadership in the Texas legislature, the governor's office, and the PUCT to persuade those parties resistant to change (both consumers and producers) that electricity restructuring was a positive-sum game worth pursuing; and in midcourse, as natural gas prices rose, that leadership had to reinforce the point, bolstered by data analysis, that it was still a positive-sum game despite the increase in input costs.

This vision and leadership was necessary to make restructuring a success, but it was not, on its own, sufficient. It had to be coupled with careful institutional design that focused on getting the rules right for promoting social welfare. This institutional design drew on the input of a wide variety of stakeholders and took into account their analyses of the design proposals. The resulting institutions, which are the primary subject of this volume, had two crucial features: They promoted clarity and transparency, and they facilitated decentralized coordination. Given the diversity of buyers and sellers in this industry with diffuse private knowledge about their own preferences and/or costs, decentralized coordination in the open market was the approach most likely to maximize welfare. The benefits that arise from it were seen in the gains from trade, or welfare creation, that accrued to all parties to this exchange. By facilitating decentralized coordination instead of imposing specific outcomes, the institutions designed in Texas became the most market-oriented in the country, and the most likely to be resilient and adaptive in the face of unknown and changing economic, technological, and environmental conditions.

Finally, once the transition started, the institutional design was not considered final. Ongoing measurement, evaluation, and analysis led to evolution of the institutions as they moved toward greater efficiency. The wholesale market design transition from zonal to nodal pricing is an ongoing example of this institutional adaptation through analysis and evaluation.

Chapter 1 presents the background of a necessary condition for Texas's restructuring: the independence of ERCOT from the Federal Energy Regulatory Commission (FERC). Authors David Spence and Darren Bush tell us that, originally, ERCOT set itself free from the federal government by refusing to allow interconnection from utilities in states outside Texas. As the electricity grid grew, and demands for electricity grew with it, isolation became less and less viable. ERCOT, however, was able to negotiate the equivalent of a treaty with FERC, persuading the federal agency to agree to forbear from exercising its authority in ERCOT. Spence and Bush explain how sole PUCT jurisdiction made it much easier for Texas to restructure effectively.

In chapter 2, Pat Wood III and Gürcan Gülen review the political and regulatory process by which restructuring took place in Texas. The 1978 Public Utility Regulatory Policies Act created the regulatory atmosphere for cogeneration of electricity by industries across Texas, especially in the highly industrialized Houston Ship Channel. The broad use of cogeneration in Texas demonstrated that electricity generation was by no means a natural monopoly, tearing down the intellectual framework for cost-based, rate-of-return regulation. Still, electricity restructuring had many hurdles to overcome after the passage of its enacting legislation, Senate Bill 7, in 1999, and the PUCT could not have made the progress that it did without substantial political support from many areas.

Once the legislature decided that wholesale electricity competition would take place, it was up to the PUCT to choose which among the competing forms of market design would be established in Texas. In chapter 3, former PUCT staffers Eric S. Schubert and Parviz Adib describe the long and tortuous journey from zonal to nodal wholesale markets. In theory, nodal markets are more efficient but require a far greater infrastructure for operation. Schubert and Adib explain how the political and economic consensus for the more expensive nodal markets was established.

Because it cannot be stored easily or cheaply, electricity supply needs to be extremely close to demand at all times. How ERCOT addressed the challenge of "resource adequacy" and kept the lights on in Texas is the subject of chapter 4, by Eric S. Schubert, Shmuel Oren, and Parviz Adib. Other system operators in the United States have sought to achieve resource adequacy by creating "capacity markets," where firms are paid for having capacity available whether it is used to generate electricity or not. The problem with capacity markets is that they may serve merely as a mechanism for generators to bring in extra monies without inducing the generation of electricity where and when it is needed or reducing demand when it is the cheapest alternative for system power balance. Schubert, Oren, and Adib explain how the PUCT and ERCOT managed to avoid capacity markets while still creating sufficient new capacity to maintain resource adequacy in the ERCOT region.

To date, there is no effective mechanism for competition in the electricity transmission sector. Ongoing technological change in distributed generation might eventually make the wires potentially competitive, but that transition has not yet occurred. Thus, continued regulation of transmission lines is required, even in a restructured electricity sector. Long-time PUCT staffer Jess Totten describes in chapter 5 the steps the PUCT has taken to ensure that the transmission sector complements wholesale market conditions. Totten explains how, moving ahead of FERC, the PUCT first reformed ERCOT and then induced ERCOT to offer open-access transmission service, allowing generators to compete on an even footing. He goes on to examine the short-run success and long-term consequences, both good and bad, of this policy.

Starting in the 1970s, it became clear that the established regulatory paradigm of natural monopoly based on central, large-scale power plants no longer applied. Restructuring can, and should, reduce entry barriers facing new, more efficient, generation technologies and business models. In chapter 6, Nat Treadway describes how smaller "distributed generation" not only adds to the supply of power in the grid, but also allows for differentiated quality of electricity supply, to the benefit of consumers as well as the system as a whole. Treadway's thesis is not that distributed generation has thrived because the PUCT allowed it to. Rather, it is that restructuring in Texas thrived in large part because the example of distributed generation predated and showed the way for restructuring. Now, in turn, the distributed generation experience in Texas can serve as a model for other states. The conduciveness of the PUCT structure in ERCOT to wholesale market competition by no means guarantees that competition is fully robust in the ERCOT market. In chapter 7, Steve Puller examines the efficiency of the Texas wholesale power market. Surprisingly, Puller finds most inefficiencies occur not because large companies exercise market power, but because small firms do not fully recognize their profit opportunities.

In chapter 8, Lynne Kiesling surveys the experience of retail restructuring and competition in Texas. In their restructuring design, Texas policymakers absorbed lessons from other states and other countries, both on what to do and what not to do. The Texas design focused on reducing entry barriers facing prospective retail competitors and implemented that concept through a "price-to-beat" mechanism that restricted the ability of incumbent retailers to lower their prices and deter others from entry. This mechanism was sufficiently flexible to adapt to an unanticipated increase in natural gas prices without harming the extent of competition in Texas markets. Among states that have implemented restructuring, Texas has had the most success in integrating wholesale and retail competition.

In the book's final chapter, Andrew Kleit reviews the PUCT's approach to monitoring the wholesale electricity market in ERCOT for anticompetitive behavior. As Kleit shows, the peculiar nature of electricity generation markets prevents competition in them from always being robust. Thus, "market monitoring," a special mechanism more intrusive than typical antitrust policy, is used to support competition.

One objective of this volume is to correct the many misunderstandings of the Texas institutional design in electricity.[1] For example, it is often argued that restructuring has led to higher electricity prices. Those who make this assertion fail to appreciate the importance of prices in transmitting information to and from consumers, especially during a time of rising fuel prices and sensitivity to environmental issues. It is true that natural gas prices more than quadrupled over a six-year period, and that a disproportionate share of generation in Texas is fueled by natural gas. But the correct comparison to make is between current retail prices and what retail prices would be right now under regulation. In 2006 the PUCT performed exactly that counterfactual analysis for the years 2002–5.[2] They found that in the Centerpoint/Reliant area (around Houston), the estimated regulated price would have been 18–26 percent higher than the average of the five lowest actual retail prices. In the TXU area (around Dallas/Fort Worth), the estimated regulated price would have been 11–18 percent higher. Although it has not been updated, this analysis (which is discussed in more detail in chapter 8) suggests that the results from retail restructuring compare favorably to those of regulation.

In other words, electricity prices would also have gone up under regulation, because the increases are driven by increases in the input price that makes up a large share of the retail price—the price of the fuel, and natural gas in particular. In competitive markets, these input cost increases are communicated more transparently to consumers, who then change their behavior, which leads to more (economically and environmentally) efficient resource use. In the case of Texas, that efficiency has taken the form of conservation and investment in energy-efficient appliances, lighting, heating, and cooling.

Second, critics contend that deregulation is to blame for consumers being switched to the legal provider of last resort (POLR) contract at high prices when their retailer chooses to exit the industry. One should remember, however, that firms commonly exit new markets (and, indeed, many established markets), and in this case those that exited the Texas market were too reliant on wholesale spot market purchases to be able to survive in a period of volatile and rising wholesale prices. Moreover, the data for Texas imply that fewer than 10 percent of suppliers exited the market. Having "quiet and seamless switching" to a POLR is one of the consumer protections embedded in the restructuring legislation, and the high retail price reflects the high degree of revenue uncertainty and wholesale market purchasing risk that the POLR faces. A POLR contract is intended to be a cushion while the customer signs up with another competitive retailer, not a market option in and of itself. From an institutional design perspective, this experience suggests that the PUCT could require more, and more timely, notification of customers prior to firm exit; it does not suggest, though, that the well-documented benefits that competition has brought to Texas consumers should be forsaken.

Finally, increased congestion, and congestion costs, in Texas during high-demand summer periods have brought complaints about restructuring. Although the nature of the problem is not now fully understood, it is clear that the increased transmission of wind power from West Texas is changing the physical flows on the network, and the outdated zonal wholesale market design (which assigns congestion costs by zone) is not flexible enough to adapt to those changes in the market and in the grid. ERCOT and the PUCT are in the middle of a multi-year process to change the market design to a nodal market, in which congestion costs are calculated more transparently.

The institutional design process in Texas continues, and it continues in a way that promises to deliver benefits both to consumers and producers through more possibilities for decentralized coordination. The PUCT and industry stakeholders are working on a framework for the (regulated) wire companies to invest in widespread implementations of advanced metering infrastructure (AMI). AMI technologies enable decentralized coordination by making possible increased two-way communication between producers and consumers. Such information flow opens up opportunities for retailers to provide more, and more varied, products and services, which they can differentiate by charging time-varying/dynamic pricing, by charging different prices for different levels of power quality, and by bundling the sale of the electricity commodity with other products and services (such as home security or home entertainment). It also creates a market for innovative digital home-automation technologies and services. Consumer value (and producer profit) are created both by providing consumers with new offerings and by making it economical and easy to take action that reduces energy use and improves environmental quality.

In retrospect, the deregulation of the airline, trucking, railroad, petroleum, and natural gas industries was relatively simple; they needed only to remove price controls and let firms compete. In electricity, however, the path of opening markets has been far more complicated. Allowing competition in generation and marketing while continuing regulation in transmission, distribution, and system operation requires thoughtful regulation at a variety of levels and an ability for the regulatory process to withstand political pressures while striving for economically efficient policies.

This task has, at least so far, proved to be beyond the grasp of state governments across the United States—except, that is, in Texas. There, focused in the regulatory island of ERCOT, the Public Utility Commission of Texas, with support from governors, legislators, and dedicated and skillful staff, has made electricity restructuring a reality. Here, then, is the Texas story—the story of how the task can be accomplished.

1

Why Does ERCOT Have Only One Regulator?

David Spence and Darren Bush

Among restructured electricity markets, Texas's market exists within a uniquely integrated regulatory environment, one in which both wholesale and retail markets are regulated by a single state overseer rather than a mix of federal and local regulators. How has this regulatory environment affected the Texas electricity market? Can one regulator regulate better than two? Can one regulator promote competition more effectively or plan better than two? Does bifurcated regulatory jurisdiction over electricity markets get in the way of efficiency? We cannot supply definitive answers to these questions here, but we hope to provide the foundation for answers by looking closely at how and why Texas has avoided federal jurisdiction over its wholesale energy markets, leaving its wholesale and retail markets in the hands of a single, state regulator.

Federal Jurisdiction and ERCOT

The story of how the Electric Reliability Council of Texas (ERCOT) has avoided most federal regulatory jurisdiction under the Federal Power Act (FPA) is less about turf battles (although it is partly that) than about political accommodation among the Federal Energy Regulatory Commission (FERC),

David Spence would like to acknowledge the research assistance of Jared Fleisher, who assisted in the preparation of this chapter.

the U.S. Congress, and ERCOT. That is, ERCOT's quasi-independence of most federal regulation is the product of litigation, legislation, and agreements.

The Setting. The story begins in 1927, before the passage of the Federal Power Act. In *Rhode Island PUC v. Attleboro Steam and Electric Co.*, the U.S. Supreme Court held that states were constitutionally prohibited under the dormant commerce clause of the U.S. Constitution from setting the price of electricity that was generated in-state but sold across state lines.[1] As states were the leading actors on the electricity regulation stage at the time, the so-called *Attleboro* gap necessitated federal intervention in the interstate energy market. Congress responded by passing the Federal Power Act of 1935,[2] which granted the Federal Power Commission (FPC, now the Federal Energy Regulatory Commission) authority over rates and conditions for the interstate sale and transmission of wholesale electricity.

The FPA extended federal jurisdiction "to the transmission of electric energy in interstate commerce and to the sale of electric energy at wholesale in interstate commerce."[3] Electric energy was "transmitted in interstate commerce" if it was "transmitted from a State and consumed at any point outside thereof."[4] At the same time, however, the law specifically denied the commission jurisdiction

> over facilities used for the generation of electric energy or over facilities used in local distribution or only for the transmission of electric energy in intrastate commerce, or over facilities for the transmission of electric energy consumed wholly by the transmitter.[5]

Since the FPA did not purport to exercise jurisdiction over wholly intrastate transactions,[6] FERC did not have jurisdiction over generating operations and "facilities used for local distribution"—that is, for subdivision and allocation to local retail customers.[7]

Thus, the boundary line between state and federal jurisdiction here was not set primarily by the Constitution's commerce clause, which as interpreted by the Supreme Court extended federal jurisdiction to activities that "substantially affect" interstate commerce. Although electricity transmission, even within a single state, "substantially affects" interstate commerce as that standard was developed by the Supreme Court,[8] statutes might exert something less than the full reach of constitutional jurisdiction. The general rule for determining the interstate character of a transmission under the FPA was the "technological transmission test," a scientific or engineering test holding that "federal jurisdiction is to follow the flow of electric energy."[9] If a transaction resulted in the movement of electricity across state lines, it was subject to FERC jurisdiction. Against this jurisdictional backdrop, ERCOT was created.

The Creation of ERCOT. The entity now known as ERCOT had its roots in the response of Texas utilities to the passage of the FPA in 1935. As Seventh Circuit Court of Appeals Judge Richard C. Cudahy described it, Texas utilities "elected to isolate their properties from interstate commerce" and to place themselves beyond the reach of the FPC, "whose jurisdiction was limited to utilities operating in interstate commerce."[10] Early on, these obligations for maintaining intrastate character were often (though not always) memorialized in informal agreements.[11] It was not until 1970 that Texas utilities formally established an intrastate power pool, the Electric Reliability Council of Texas, or ERCOT.

As formed in 1970, ERCOT was a "regional electric reliability council," one of many overseen by the North American Electric Reliability Council. Thus, it was not an entity exercising delegated state power, and it was not established by state-enabling legislation. Indeed, no state electricity regulation existed in Texas until 1975. Rather, ERCOT was formed as a voluntary membership organization. Texas utility representatives understood that a condition of such membership was the requirement of any entity making interstate connections to advise the other members, which would then sever their connections with that entity to maintain ERCOT's intrastate character.[12]

Over time, ERCOT evolved to take on formal responsibility for overseeing the reliability of wholesale and retail electricity markets, as authorized by state statute and overseen by the Public Utility Commission of Texas (PUCT). In 1975, the Texas legislature initiated state regulation of the energy market with the Public Utility Regulatory Act (PURA),[13] granting the PUCT traditional state regulatory powers, including ratemaking. ERCOT continued to operate as a private coordinating council in this newly regulated market.[14]

Thus, ERCOT was a response to the technological transmission test: a way to ensure that the flow of electrical energy within the ERCOT system was contained within the state of Texas.[15] As the other lower forty-eight states operate upon a grid interconnected among many states, and as the physical properties of an electrical system mean that a single interconnection will instantaneously link whole systems across state lines, the test has resulted in an expansive role for federal jurisdiction everywhere but Texas.[16]

ERCOT's Jurisdiction in Modern Restructured Electricity Markets

Although ERCOT's jurisdictional autonomy was originally founded upon its lack of interconnection and transmission across state lines, it is nevertheless linked today by asynchronous connections to the Southern Power Pool (SPP) in Oklahoma. Two such interconnections were established in 1981 and 1983 as part of FERC-approved settlements to a lengthy jurisdictional controversy. Combined with the gradual restructuring of the electricity industry over the past three decades, the links have given rise to litigation over the jurisdictional boundary between FERC and ERCOT.

Testing ERCOT's Independence: The CSW Cases. The first wave of litigation arose with respect to a holding company, Central Southwest Holdings (CSW), which had utilities in both Texas and Oklahoma. To comply with provisions of the (now repealed) Public Utilities Holding Company Act (PUHCA) that required integration of its far-flung facilities, CSW needed to establish an electrical connection among its four constituent utilities, which it could not do without violating ERCOT's intrastate character. In 1974, a group of utilities in Oklahoma forced the issue by filing an action asserting PUHCA noncompliance by CSW.[17] As a first effort to force interconnection, CSW mounted an antitrust challenge alleging that ERCOT amounted to an illegal restraint of trade in interstate commerce; this argument was rejected by the courts on the grounds that intrastate isolation was a permissible end under the Federal Power Act, and that the plaintiff failed to establish the existence of any unlawful restraint of trade, conspiracy in restraint of trade, or unlawful boycott.[18]

CSW then sought to direct a power flow originating within ERCOT into the SPP in Oklahoma. On May 4, 1976, one of the CSW companies began transmitting power from its southern division in Texas for resale to customers in Oklahoma. This became known as the "midnight connection." CSW's intention was to subject ERCOT to federal jurisdiction under the "technological transmission test," and then petition FERC (at that time the FPC) to order interconnection.[19] Pursuant to ERCOT's charter, ERCOT members disconnected from the CSW system.

CSW's petition to the FPC was, however, filed prior to the disconnection, when the defendant ERCOT utilities were still transmitting in interstate commerce. (The case was decided after such transmissions had been disconnected and isolated.) The FPC held CSW's constituent utilities to be "public utilities" subject to federal regulation due to their interstate connection, but found that the other ERCOT utilities, which had disconnected and formed a bifurcated system shortly after the midnight connection, could not be subjected to federal jurisdiction, presumably because their operations, transmissions, and interconnections were now wholly intrastate. In the absence of a finding of jurisdiction, the FPC rejected CSW's claims for forced interconnection, although it ordered Texas utilities to restore certain connections with CSW for reliability reasons, explaining that connections established only for emergency purposes would not trigger federal jurisdiction.[20] Subsequently, at the urging of ERCOT utilities, the PUCT ordered all connections with CSW severed, and this order was upheld by the courts.[21]

The passage of the federal Public Utility Regulatory Policies Act of 1978 (PURPA)[22] endorsed (and broadened) the compromise represented by the FPC order in the CSW cases. Specifically, PURPA authorized FERC to order interconnection and wheeling in Texas,[23] including interconnection to interstate lines, without triggering general FERC ratemaking jurisdiction over transactions taking place strictly within the state.[24] These changes ultimately induced CSW to settle its claims against ERCOT. On July 28, 1980, both CSW and the other ERCOT utilities submitted a settlement proposal that sought "approval of two asynchronous direct current interconnections between electric utilities in ERCOT and [the Southwest Power Pool, in Oklahoma]."[25] FERC accepted the settlement offer,[26] under which it ordered interconnection and transmission without subjecting ERCOT to plenary federal jurisdiction.[27]

ERCOT's Jurisdiction in the Age of Restructuring. The restructuring of the electricity industry following passage of the Energy Policy Act of 1992 (EPAct of 1992)[28] accommodated the arrangement for jurisdiction-sharing between ERCOT and FERC. EPAct of 1992 was passed to "open and expand the wholesale transmission market and encourage the development of new competitive generating companies," and it armed FERC with additional powers to do so.[29] It amended the FPA to enhance FERC authority such that "any electric utility . . . or any other person generating electric energy for resale, may apply to the commission for an order . . . requiring a transmitting utility to provide transmission services to the applicant."[30] EPAct of 1992 also added a separate provision for ERCOT facilities, however, effectively reserving ratemaking jurisdiction for transmission services within ERCOT to the PUCT rather than FERC.[31]

Shortly after the passage of EPAct of 1992, FERC issued Order No. 888, mandating open-access wholesale wheeling at nondiscriminatory rates and the unbundling of wholesale sales from transmission services.[32] Importantly, Order 888 did not reach ERCOT, referring as it did to "public utilities" and concerning itself with ratemaking (which EPAct of 1992 had left to ERCOT within Texas). FERC reserved the right, however, to order wheeling by electric utilities, including ERCOT utilities, under sections 211 and 212 of the FPA.[33]

By this time, Texas was restructuring its electricity markets, as well. In the mid-1990s, amendments to the PURA deregulated the Texas wholesale market, and, in 1996, the PUCT employed its rulemaking authority under these amendments to make ERCOT the first "independent system operator" (ISO) in the United States, thereby giving ERCOT "responsibilities over wholesale competition and for ensuring efficient use of the transmission network by all market participants."[34] In 1999, the Texas legislature passed Senate Bill 7, which created a competitive retail market and granted the PUCT the authority to certify ERCOT as the independent organization overseeing network reliability and retail operations.[35] Since the passage of Senate Bill 7, the PUCT and ERCOT have overseen the transition to competitive electricity markets in Texas.[36] Indeed, FERC recently approved a third asynchronous connection with the interstate grid, without creating any threat to sole PUCT jurisdiction over ERCOT.[37]

Tensions between FERC and the States

With Order No. 888, FERC effectively unbundled wholesale electricity sales from transmission services, opening the generation and wholesale electricity markets to competition. Shortly thereafter, FERC began to grant widespread permission to wholesalers of electricity to base their sales on market rates. Meanwhile, a similar process was under way in the states, where the large retail customers of monopoly investor-owned utilities (IOUs) pushed for the right to choose their electricity suppliers,[38] urging state legislatures to open retail markets to consumer choice. California, New York, Pennsylvania, Massachusetts, and several other states—notably, those with relatively high retail rates—obliged during the mid- and late 1990s by opening their retail markets to competition.[39] Although its retail rates were not particularly high compared to the national average, Texas, too, joined the group of restructuring states.

While all these states sought to replace administratively set retail prices with prices set by the market, retail restructuring necessitated policy decisions about wholesale markets, as well. If retailers (IOUs and their new competitors) could not remain fully vertically integrated, they would have to acquire their power on wholesale markets; sometimes, however, restructuring states imposed restrictions on their wholesale market transactions. Following the lead of the United Kingdom's electricity market, California in its restructuring plan adopted the so-called poolco model, requiring retailers to acquire all their power through the California Power Exchange, a clearinghouse to which buyers and sellers of wholesale power submitted daily bids for short-term power transactions. By contrast, Texas and many of the other states chose to allow retailers to rely on long-term bilateral contracts, and to use short-term markets to supplement retailers' needs and to balance loads. Similarly, while many other states created retail price *ceilings* during the transition to competition to protect consumers from price spikes, Texas imposed a retail price *floor* on its incumbent IOUs to foster market entry.[40]

Despite these differences among approaches to retail restructuring, the sudden and drastic price increases on the California wholesale market in the winter of 2000–1 seemed everywhere to undermine faith in the idea that restructuring would bring lower prices. Historically, wholesale prices had hovered at or below $50 per megawatt-hour (MWh) in California; but in December 2000, prices on the California wholesale electricity spot markets reached monthly averages of nearly $400/MWh, while some daily prices averaged nearly $1,200/MWh.[41] While a supply-demand imbalance and cost factors were the largest parts of the problem, regulators and academics subsequently determined that the state's poorly designed market also created easy opportunities for sellers to exert market power over price and to otherwise game the system.[42] Many sellers took these opportunities, provoking enforcement actions by FERC, the U.S. Commodities Futures Trading Commission, and the U.S. Department of Justice that resulted in more than 3 billion dollars in refunds, hundreds of millions of dollars in civil and criminal fines, and prison sentences for some of the individuals involved in the scandal.[43]

During the crisis, the responses of FERC and California regulators were not well coordinated, to say the least. Both the inconsistency between relatively free wholesale markets and retail price caps and California's unhappiness with FERC's initial responses to soaring wholesale prices testified to that coordination problem. As electricity shortages developed in California during the winter of 2000–1, FERC's preference for letting wholesale prices rise (to encourage more investment in supply and discourage consumption) conflicted with the state's policy of capping retail rates, which discouraged both investment in supply and conservation. In a desperate attempt to reduce price volatility, California asked FERC to implement a series of wholesale price caps,[44] designed to mitigate generators' market power by capping wholesale rates they would receive in the real-time market. Because the caps were at first implemented only in California, however, they were largely unsuccessful,[45] as generators began selling to points outside of the state, where the wholesale price was still volatile and not subject to a price cap.[46]

Meanwhile, the California ISO purchased power from out of state at uncapped rates because it needed it desperately. Some generators took direct advantage of the arbitrage opportunities posed by the differing rates in the western markets by exporting power outside the state and shipping it back in at uncapped rates.[47] In these ways, the lack of a single market with a single regulator encouraged arbitrage, hindered the operation of the California market, and facilitated the exercise of market power by sellers in it.[48]

Other wholesale markets besides California have had to confront the coordination issues that emerge from dueling federal and state regulation.

By securing capacity and energy from beyond the jurisdictional reach of their governing ISOs, they have created a new kind of regulatory gap, sixty years after the closure of the *Attleboro* gap—one that seems to have exacerbated the breakdown of markets in California. FERC, recognizing that nondiscriminatory retail wheeling was a prerequisite to efficient markets, anticipated this coordination problem by asserting jurisdiction over retail wheeling in Order No. 888.[49] Order 888 did not close the gap completely, however, as the California experience illustrates.

Since the California crisis, FERC's oversight of wholesale power markets has evolved. Initially, FERC gave states and ISOs a great deal of leeway in the design of wholesale power markets. After the crisis, FERC oversight became more stringent and moved toward greater federal intervention into wholesale power markets, particularly to guard against the rise of market power and its misuse. Antitrust and public utility laws are two means by which FERC has strengthened its oversight. We turn to that subject next.

Market Power, Antitrust Law, and the Filed-Rate Doctrine

The introduction of competition into formerly regulated markets has focused increasing attention on the relationship between antitrust law and restructured (but still regulated) electricity markets, while the California crisis has exacerbated fears that markets are vulnerable to manipulation. The Federal Power Act continues to require wholesale rates to be "just and reasonable," a mandate FERC must enforce. Similarly, Texas law requires electricity to be "reasonably priced," a mandate the PUCT enforces. These mandates continue to have the force of the law, even though prices in Texas and other restructured markets are now set by market forces rather than by regulators.

Outside of such regulated industries as electricity, the primary instrument for managing competition is antitrust law; however, because the electricity industry has been subject to intrusive regulation (including price regulation) for most of its existence, courts have historically employed a number of doctrinal devices by which they reserve to energy regulators the job of policing competition. One such doctrinal device is the "filed rate," or the *Keogh* doctrine (named after the case that established the principle), which bars private plaintiffs from recovering damages for allegedly anticompetitive electricity rates charged under tariffs that have been filed with regulators.[50]

The *Keogh* doctrine was established in an era of administratively approved rates represented in tariff schedules filed with regulatory commissions. The rationale for the rule is that private antitrust enforcement makes no sense when a regulatory agency charged with ensuring reasonable prices sets the price and the terms of competition. The doctrine has cracks, however, through which anticompetitive behavior can slip. Even under rate regulation, the regulatory agency may not have the general power to determine whether any particular conduct, beyond the rate itself, is anticompetitive.[51] Furthermore, in restructured, competitive markets, the determination of the justness and reasonableness of a rate must be made *ex post* rather than before tariffs are filed.

Nevertheless, the doctrine is now typically applied to rates set in deregulated markets that are subject to some degree of price competition, despite the fact that the rate charged—the market price—is not really "filed," but, rather, is set by market forces. Much of the case law extending the filed-rate doctrine to competitive electricity markets arose out of the "manipulation" of California's deregulated electricity market.[52] In that context, the Ninth Circuit Court of Appeals has consistently applied (some might say, misapplied) the doctrine to prices set in a market. For example, in *Public Utility District No. 1 of Snohomish County v. Dynegy Power Marketing Inc.*, the Ninth Circuit, citing the filed-rate doctrine, upheld dismissal of a claim brought under state antitrust law seeking monetary and injunctive relief for rates alleged to have been manipulated in California's electricity market.[53] In support of this conclusion, the court cited FERC's extensive oversight of rates in the California market, including things like

- FERC's grant to Dynegy Power Marketing of the authority to engage in market-based ratemaking;
- the requirement that firms operating under market-based rates file regular reports with FERC "to ensure that rates will be on file as required by [the Federal Power Act], to allow FERC to evaluate the reasonableness of the charges as required by [the Federal

Power Act], and to allow FERC to continually monitor the seller's ability to exercise market power"; and

- FERC's approval of the market rules for California's electricity market.[54]

The court also noted that FERC had ordered disgorgement of profits by the energy marketers for the anticompetitive practices alleged by Snohomish County.[55]

The Ninth Circuit's analysis, however, was problematic. The market structure that FERC approved, the basis of the rates sought, and, indeed, the authority of the firms to act within that market were all based on the notion that the market would be competitive, that prices would be determined based on the marginal cost of producing the last unit of energy in each hour of the day, and that excess capacity would serve to discipline price.[56] What occurred instead was the now familiar broken market in which owners of generation had both the incentive and ability to exercise market power. Consequently, prices in California markets were well above competitive levels, and they could not move toward competitive levels, since demand (under capped retail prices) could not respond to wholesale price increases. While the wholesale prices were covered by pro forma market-based rate filings at FERC, under the Federal Power Act FERC could only alter those prices *ex post*, after a finding that they were unjust and unreasonable. FERC's only remedy in that instance was disgorgement of unjust profits, a far less effective deterrent than treble damages in a private antitrust action.[57]

Does the filed rate doctrine apply to barring private suits in the wholly intrastate Texas electricity market? That question was addressed in the recent case of *Texas Commercial Energy L.L.C. v. TXU Energy Inc.*,[58] in which the Fifth Circuit Court of Appeals rejected the claim that a market-based rate was not a filed rate and concluded that the *Keogh* doctrine applied equally to rates filed before a state commission. Texas Commercial Energy argued that the Texas statute allowed private antitrust claims even if the Federal Power Act did not,[59] but the Fifth Circuit disagreed. The court held that "applying the filed rate doctrine, along with other common-law defenses that are normally part of the federal antitrust legal landscape, gives effect to the [Texas] legislature's intent" in enacting its public utility law.[60]

Because electricity markets are still very young, so, too, is litigation exploring the continued vitality of doctrines protecting electricity firms from some forms of antitrust law. For now, both in Texas and elsewhere, courts continue to apply the filed-rate doctrine to protect electricity firms from private antitrust enforcement.

Market Power and Capacity Assurance

Generators and wholesale sellers were able to exercise market power in California in part because of the imbalance between supply and demand. For that reason, regulators worry about ensuring an adequate supply of energy and transmission capacity over the long run. In competitive markets, regulators encourage ISOs to promote investment, and ISOs employ a mix of mandates and incentives to ensure that energy is not too scarce. Some try to price energy in ways that will offer incentives for construction of new capacity. This is the route Texas has chosen.[61] Others require retail sellers of energy in competitive markets to maintain adequate reserves to satisfy peak demand through the use of devices such as capacity markets.[62] Others organize the acquisition of new capacity through auctions.[63]

The problem of encouraging development of capacity elsewhere, however, is complicated by FERC's lack of authority to oversee siting of network facilities. While pricing regimes can offer strong incentives to build new lines or generation units, they cannot ensure the siting of either. The law gives the states (rather than FERC) the authority to approve, or to veto, construction of new capacity. Communities that will not benefit from the presence of new network lines (because they neither sell energy into nor buy energy from the line) may choose to withhold that permission. So FERC's attempts to encourage investment (through ISOs) depend upon the will of the states to allow new capacity to be built.[64]

Providing prospective investors with incentives to invest in new capacity is tricky everywhere. All states enjoy the power to regulate the process of siting new generating capacity and transmission lines—a combined jurisdiction that presumably greatly simplifies (and rationalizes) the capacity-planning process. The difference for Texas is that, while other ISOs derive their power over grid owners and users from FERC, which has little power over siting, ERCOT derives its power from the more influential PUCT. ERCOT has not been immune from the problem of ensuring adequate capacity, but the process of finding a solution is not further complicated by jurisdictional coordination issues. This may be one reason initial fears about resource adequacy in Texas have been quelled in recent years.[65]

Conclusion

Emerging electricity markets have experienced growing pains, and Texas's market is no exception. For regulators everywhere, the task of fine-tuning rules to promote competition, prevent and address abuses of market power, and provide an incentive for investment in capacity is no small feat. Both FERC and the PUCT have tried to accomplish this challenging task, with differing levels of success. The decision to create ERCOT to avoid the kinds of interstate transactions that subject it to FERC jurisdiction has undoubtedly simplified the regulatory task in Texas. Certainly, it is theoretically possible that state and federal regulators could coordinate their efforts elsewhere in ways that promote the development of markets efficiently. The process of helping these markets reach maturity is, however, a delicate one that will take time. The PUCT and ERCOT seem to be working effectively together to promote competition in Texas electricity markets, and ERCOT will continue to enjoy its exemption from FERC jurisdiction for the foreseeable future. While we do not yet know the connection between regulatory integration and market performance, we can speculate that the relative competitiveness of the ERCOT market may be due in part to the lack of split regulatory jurisdiction over wholesale and retail markets.

2

Laying the Groundwork for Power Competition in Texas

Pat Wood III and Gürcan Gülen

Since the late 1980s, many jurisdictions around the world have experimented with restructuring their electricity markets. Many have failed. As a result, the topic of competitive electricity markets has become politically difficult in the United States, especially since the California crisis of 2000–1. Unlike other states, however, Texas has moved forward with reforms and today has one of the most vibrant competitive electricity markets in the world, in both the wholesale and retail sectors. In this chapter, we provide a historical perspective on how Texas restructured its market, and look in particular at how it balanced the interests of various stakeholders and focused on economic fundamentals.

Prelude to Restructuring

During the 1950s and 1960s, mainly due to large numbers of first-time air conditioning installations, growth in electric power demand in Texas averaged 10 percent per year. Inexpensive fuel and large power plants held prices to two cents per kilowatt-hour for residential customers. Nationally, this rapid growth of electricity consumption continued until the early 1970s, when it was curtailed by high fuel prices, environmental concerns, and a slowing U.S. economy. While most of the United States struggled with rising fuel costs, Texas was blessed with an abundance of locally sourced oil, lignite, and, especially, natural gas, which kept prices low.

During the 1960s, natural gas had become the fuel of choice for electricity generation in Texas. But in the 1970s, bifurcated federal-state regulation led to pronounced imbalances between the supply and demand of natural gas. Laws passed that decade, including the federal Powerplant and Industrial Fuel Use Act of 1978 (PIFUA),[1] prohibited the construction of new gas-fired baseload power plants. To meet growing power demand, utilities began to look to other fuels. Plentiful local sources of lignite led to development of mine-mouth coal-fired generation, but with the cost of coal shipments from other parts of the United States prohibitive, Texas turned next to nuclear power. Two nuclear facilities were built in the state: the South Texas Nuclear Project (by Houston Lighting & Power) and the Comanche Peak nuclear facility (by Texas Utilities).

To cover the cost of building these coal and nuclear facilities, utilities asked for and received rate increases. Customer dissatisfaction in response to some increases of more than 50 percent eventually led to greater scrutiny by regulators, who began to withhold full approval in some cases.[2] Most of the ratepayer litigation relating to these regulatory cost-increase decisions was settled by the 1990s. By that time, the coal and nuclear plants, which took from five to fourteen years to build, were all operational; but the growth in electricity consumption in Texas had declined sharply from earlier projections. Thus, by the early 1990s, Texas had substantial excess baseload capacity.

Two crucial policy changes at the federal level had an initially subtle, but ultimately profound, impact on the Texas power marketplace. The first was the Public Utility Regulatory Policies Act (PURPA) of 1978.[3] Two primary objectives of PURPA were, first, to encourage energy efficiency through expanded use of cogeneration (the simultaneous generation of power and heat or steam at industrial facilities), and, second, to promote the use of renewable fuels and fuel wastes. PURPA required utilities to purchase electricity produced by cogeneration and from small power-production facilities (known as "qualifying facilities," or QFs) that used renewable fuels at avoided cost—that is, costs the utilities would have incurred if they had constructed new facilities and generated the electricity themselves. QFs could use some of the electricity they produced to serve the electrical load of a host industrial or commercial facility, such as a refinery, and sell the excess back to the utility; they could also sell their total output to the local utility. In addition to power,

QFs typically produced steam, which was sold to the host facility. QF owners were given exemptions to the Public Utility Holding Company Act of 1935 (PUHCA),[4] making it possible for a large number of companies that were not utilities to enter the electricity generation business.

The other crucial federal reform of 1978 was the Natural Gas Policy Act (NGPA).[5] The NGPA was intended to remove the distortions associated with federal control of wellhead natural gas prices, which most now agree was the cause of the natural gas shortages in the 1970s; its goal was to end federal price controls by 1985. The most controversial part of the National Energy Act, the NGPA was very complex and difficult to implement. Phasing out price controls resulted in increased upstream activity, which revealed that there was plenty of natural gas in the United States. Production surpassed demand, creating a "bubble" and depressing prices several years before the 1985 target. The bubble persisted through the 1990s as new resources were developed and natural gas was supplied to generators around the country, especially in producing areas such as Texas.

How did these federal reforms play such a large role in the development of competitive power markets in Texas? After all, the Texas intrastate interconnected electric system (known as ERCOT, the Electric Reliability Council of Texas) was basically its own economic and jurisdictional island. The answer is simple: In all of the United States, there was no greater concentration of cogeneration than the highly industrialized Houston Ship Channel, the heart of the service area of Houston Lighting & Power.[6] By tipping the scale in favor of nontraditional sources, PURPA proved that power generation was not a natural monopoly, and thus set the stage for competitive power generation, especially using the natural gas that had become available at economical prices.

Nonutility power generators faced little competition from new utility generation. The PIFUA prohibited utilities from using natural gas as a boiler fuel in their own facilities. But the fuel of choice for almost all cogeneration was gas, and power generated in the highly efficient cogenerated facilities from plentiful, inexpensive natural gas was the only economical source of new power that regulators would countenance. In the 1980s and early 1990s, customer discontent at marked increases in the price of regulated power was a front-burner issue at the Public Utility Commission of Texas (PUCT). Thus, a utility with a retail service obligation and thrifty regulators had only one way to meet its needs: buy gas-fired power from nonutilities.

Between the federal reforms of 1978 and the mid-1990s, the ingredients for profound change came together in Texas: high average costs for power due to the inclusion of expensive new coal and nuclear power stations, low incremental costs of new gas-fired power, and a large class of customers who had (and used) the ability to generate their own power. This supply and cost environment joined a policy environment characterized by a growing national consensus toward more competitive energy markets, a business-oriented political climate in Texas that focused relentlessly on economic development, and the election of a new governor in 1994 with a strong belief in markets forged during his years in the competitive West Texas natural gas business.

1995 Legislative Changes

Even before the dust of the 1994 Texas election of Governor George W. Bush had settled, all the players in the energy marketplace knew that the 1995 biennial session of the legislature was going to be the electricity battle royal. The reauthorization of the PUCT and its governing statutes pursuant to Texas sunset laws, which had been on deck in the 1993 legislative session, was aborted that year in a high-profile, last-minute scuffle over the "phantom tax," a ratemaking issue relating to the reimbursement of utility income tax. This left the entire Public Utility Regulatory Act (PURA) wide open for revisiting in 1995.

Meanwhile, the Federal Energy Regulatory Commission (FERC), armed with new powers under the federal Energy Policy Act of 1992, was beginning to restructure interstate power markets. A group of rural cooperatives in eastern Texas had submitted a petition to FERC seeking open access to the transmission grid of Texas utilities not regulated by the agency[7]—a most unwelcome event in Austin, as the ERCOT utilities, the Texas legislature, and the PUCT feared federal regulatory intrusion into a state-regulated domain. Large industrial customers, particularly those with multiple plants in Texas, saw the 1995 session as their best opportunity to enact "retail wheeling"—that is, to use their cogenerated power at multiple locations without depending on the purchase of more expensive system supply from local regulated utilities.

Retail wheeling was alarming, however, to the large investor-owned utilities, or IOUs (HL&P, Texas Utilities, and Central Power & Light) that had only recently achieved the full inclusion of their large coal and nuclear investments in regulated retail rates. Adding to their anxiety was the skepticism with which the PUCT was viewing special "economic development" rates for large industrials, a program used successfully in other states to eliminate the agitation for retail wheeling. On December 1, 1994, FERC issued an order in Docket No. TX94-4 granting the East Texas Electric Cooperatives federal open access to Texas Utilities' transmission system, threatening the state's independent control over its own transmission grid and further intensifying these concerns.[8]

In late spring 1995, as negotiations over the PUCT sunset legislation were at a height, the Texas Supreme Court resolved the 1993 "phantom tax" issue in the utilities' favor. This action took the issue off the table, allowing the parties to focus on the core debate over market design. From this legislative cauldron, a robust wholesale competition regime emerged as the workable compromise between a rudderless status quo and an unpredictable new regime of retail competition, made possible by the passage on May 1995 of Senate Bill 373.[9]

In restructuring the Texas wholesale power market, the first step was to remove barriers for new entrants. SB 373 permitted participation of exempt wholesale generators (EWGs), a federal classification that allowed the exemption of independent power producers from sections of the onerous PUHCA. Power marketers and EWGs were not defined as utilities in PURA but were authorized by the new law to sell only wholesale electric power, thus exempting them from onerous state utility regulation as well. They could be affiliated with, and sell power to, a public utility, but they had to register with the PUCT.

SB 373 also required utilities to provide unbundled wholesale transmission service at rates, terms, and conditions comparable to their own use of their systems, a step that FERC later took in its landmark Order No. 888 (April 24, 1996).[10] In a parallel to FERC's section 211 power, which was used in the East Texas Cooperatives case mentioned above, the PUCT was given power to require a utility to provide transmission access to another utility, a power marketer, a QF, or an EWG.

Wholesale power rates were not deregulated in 1995 except for those of river authorities. Utilities were allowed to request approval to discount wholesale tariffs below their approved rates. Similarly, retail utilities were given clear authority to offer flexible retail pricing, with retail tariffs or contracts that were less than their approved rates. These rates had to be greater than the utilities' marginal costs, as approved by their regulatory authority (either the PUCT or municipal government). In implementing both of these provisions, however, the PUCT ruled that the costs for these discounted rates could not be shifted to other customers, motivating industrial customers to continue to seek broader changes to the retail regulatory regime.[11]

Although the 1995 omnibus restructuring bill laid the foundation for more competitive markets, it also included an integrated resource plan (IRP) or least-cost IRP provision, drawn from the failed 1993 legislative session negotiations, which eschewed a free-market approach. Instead, it called for the PUCT to be heavily involved in decisions about utility resource procurement in order to ensure that "prudent" investments were made. Prudence reviews were often used for regulatory oversight. The IRP approach evolved across the states, including Texas, in response to the large costs associated with new additions to generation capacity in the 1980s, especially nuclear plants. Under SB 373, starting in 1996, utilities were required to file an IRP every three years. Among many other provisions was a requirement for customer participation in developing each utility's IRP—a requirement that had a most interesting consequence for the development of renewable energy, as will be discussed later.

Implementing Wholesale Competition

Immediately following the enactment of SB 373, the PUCT opened a rulemaking,[12] which became the vehicle to establish a competitive wholesale market in ERCOT. In the summer of 1995, PUCT chairman Pat Wood proposed to replicate in the ERCOT power market the natural gas wholesale restructuring approach of FERC Order No. 636[13] and lay the groundwork for retail competition in the future. His "straw-man" proposal entailed functionally unbundling the vertically integrated utilities, redefining the obligation to serve, calculating and beginning recovery of generation costs that could be stranded, and creating genuine open access, facilitated by an independent grid operator. In the end, only the last item was pursued by the

PUCT in 1995, but, with clear statutory authority and swiftly promulgated rules that opened up access on the state's transmission grid to all wholesale buyers and sellers, Texas was on its way to competition.

Because there was little political support in the state for transmission ownership divestiture, the PUCT publicly supported an entity that would ensure independent operation of the power grid. In early 1996, ERCOT, the reliability council overseeing the intrastate grid, was prodded to reinvent itself as a wholesale independent system operator (ISO). The "reinvention" of ERCOT was not an easy sell, however. Some members of the existing organization were less than enthusiastic about competitive power markets and did not want to alter the status quo that ERCOT represented. Others viewed ERCOT as a "utility-only club" that could not be trusted to facilitate competitive markets.

Despite these objections, most parties saw the inevitability of market changes and brokered a deal on ERCOT board composition, staffing, and responsibilities that comported with the PUCT's stated goals. An ongoing nonpublic investigation by the Antitrust Division of the U.S. Department of Justice into possible discriminatory practices at ERCOT and the major Texas utilities also contributed to the constructive environment. In the late spring of 1996, the PUCT approved the consensus proposal morphing ERCOT into a broad-based, independent organization, and the nation had its first ISO, which became operational in July 1997. At the time, the process that led to the re-creation of ERCOT—one driven by consensus among market participants—was taken for granted, but in the years to come, it came to be viewed as the key to Texas's success.

In addition to the swift implementation of transmission open access, the PUCT drove the development of ERCOT as a one-stop shop where wholesale customers could sign up for (and receive) transmission service, and where generators could obtain interconnection under a standard agreement. Removal of barriers to entry into the wholesale market was the key objective, and its attainment colored many decisions during this period.

Legislative Activity in 1997 and 1999: Senate Bill 7

In 1997, the Texas legislature considered a number of electricity restructuring bills. Following an adverse PUCT rate decision for the state's third-largest utility, Central Power & Light,[14] IOUs decided to support a bill introduced midway through the session, which was also supported by Governor Bush and the PUCT. Although the bill died for lack of support from other affected parties in the power market, the acute interest it provoked spurred the creation of the seven-member Texas Senate Interim Committee on Electric Utility Restructuring to study issues and report back to the Senate before the next legislative session started in January 1999. The committee conducted hearings around the state to gather information and understand better the concerns of various stakeholders. It also visited other states and countries that were restructuring their electricity markets. Another group, the Texas House State Affairs Committee, studied restructuring issues and held hearings as well, often jointly with the Senate Interim Committee.

These interim study activities were vital in teaching policymakers about the core issues of restructuring. Because regulators and lawmakers discussed and heard—together—the pluses and minuses of restructuring efforts elsewhere, they developed a working relationship outside the bright lights of hearing rooms and the heated cauldron of a four-month biennial legislature. This forging of trust between legislative leadership and regulators, uncommon in Texas, gave political cover to the PUCT for the hard calls it would be making in the years ahead.

Considerations behind Senate Bill 7. It is remarked in Austin that every major bill takes two sessions to achieve (lobbyists more crassly, and more accurately, call them "two-Lexus issues"). Senate Bill 7, the comprehensive legislation that effected electricity restructuring, took three sessions.[15] It was enacted by overwhelming bipartisan majorities of both the Texas House and Senate and signed by Governor Bush in June 1999. The law set January 1, 2002, as the starting date for retail competition.

One of the biggest concerns expressed to the interim committees was whether small (mainly residential) customers would be able to benefit from restructuring. Like many elected officials, members of the Texas legislature had a strong political interest in promising, and even guaranteeing, that restructuring and competition would lead to lower prices. To secure immediate benefit to small users (that is, voters), lawmakers required a 6 percent reduction from the end-of-1999 rate for the "safe harbor" default service for residential and small commercial customers when retail competition began in

2002. Large commercial and industrial customers had no default rate; when the gates opened, they were on their own to negotiate for their power services. Consequently, the market for these customer classes developed quickly and deeply, as they represented more than one-third of the state's load. For the residential and smaller commercial customers, though, service at the onset of competition would be provided by the affiliated retailer of their historic utility at the 6 percent discounted price, called the "price to beat." This concept, inspired by a December 1998 visit by the Senate Interim Committee with Pennsylvania power restructuring stakeholders in Harrisburg, would be valid for five years or until the affiliated retailer lost 40 percent of its customers to other providers.

One aspect of the Texas "price to beat" not present in other states' programs for default provider rates was a fuel tracker, which allowed the affiliated retailer to reset rates up to twice annually during the "price-to-beat" period to reflect changes in fuel prices. (When we consider that the later bankruptcy of California's PG&E and severe financial problems of SoCal Edison and San Diego stemmed from their inability to track wholesale prices in their retail rates in 2000 and 2001, this provision seems almost prescient.[16]) For the Texas-affiliated retailers, the provision proved fortuitous when gas prices spiked in the fall of 2005 after hurricanes Katrina and Rita disrupted natural gas markets and sent the price of Texas's dominant fuel source into the $14/Mcf range. Because it was a one-way (upward) ratchet, though, the affiliated retailers were not required to lower those prices when natural gas prices fell again. This asymmetry led to a consumer backlash, which would have been stronger had competitive retailers (and political leaders) not used it as an opportunity to persuade customers to switch providers and thereby discipline the retail market. As of the summer of 2007, six months after the final rate-cap protections expired, over two-thirds of small customers had selected competitive service, mostly from competitive providers, but also from lower-priced programs offered by the affiliated retailer.[17] By then, some retail offerings were pricing service *below* the last regulated rate in 2001,[18] even though natural gas prices had settled at a level three times what they had been six years earlier.

Texas's utilities, which had only in the early 1990s been able to roll the costs of their large generation investments into the rate base, wanted to make sure they would be able to recover these investments (made under a regulated return regime) once customers had the ability to choose other retail providers. In the end, SB 7 included a competition transition charge (CTC) to allow utilities to recover all of their net, verifiable, nonmitigated stranded costs—that is, power plant costs that had been deemed prudent by regulators before retail competition was introduced but were now above the market value of those plants. The bill defined varying quantification methods for stranded costs, based on market valuations to be calculated after the market had opened up.

To qualify for full recovery, utilities were required to clean up their power plants to a much stricter standard. These plants, mostly coal-fired, had been grandfathered from compliance with the federal Clean Air Act. Aware of the utilities' ability to use the legal system to impede regulatory mandates, Governor Bush strongly supported a voluntary approach to cleaning up emissions, and this *quid pro quo* provision of the electricity restructuring law fit his objectives. This deal to recover stranded costs for plant cleanup resulted in most of Texas's coal-fired plants coming into compliance by 2005 with the emissions standards then current.

TXU (formerly Texas Utilities) ended up achieving a global settlement of all of its transition-cost issues, and the other four IOUs litigated theirs.[19] Although the final stranded-cost calculations (known as "true-up") in 2004 were tedious and contentious, they did result in quantification and securitization of some $5 billion in excess generation costs over market. These obligations were to be recovered over fourteen-year periods by each of the affiliated distribution companies through securitized wires charges billed to retailers.

Not all utilities in ERCOT were made subject to all the SB 7 provisions. Many small cooperatives and municipal utilities (which collectively served about one-sixth of the ERCOT retail load) were concerned about their ability to compete in the new market. Losing their members, especially the largest ones, to competitive suppliers could have undermined the economic viability of their utilities and put at risk service to smaller members, who might not have been able to find willing competitive suppliers. In a pragmatic compromise, SB 7 gave cooperatives and municipal utilities the right not to join the competitive market. As of early 2008, only Nueces Electric Cooperative had joined, but its relatively successful market entry was leading others to consider opting in as well.

A related concern for independent producers, power marketers, customers, and policymakers was the ability of incumbent utilities' generation and retailing units that were to be deregulated to exert market power. This concern was addressed in SB 7 in several ways. The law required cost and corporate unbundling of generation and retailing from the regulated wires business, and the commission was required simultaneously to design expedited rate-case procedures (with a unique "future test year" feature) and pure transmission and distribution rates in the competitive market. As outlined earlier, the affiliated retailer was restricted to charging the retail "price to beat," with no ability to discount selectively to retain certain customers. Also, generators were restricted to holding no more than 20 percent of installed capacity within ERCOT and required to auction off a portion of their capacity each quarter, allowing smaller retailers to have access to generation at a market price.

At the time SB 7 was considered, a good deal of debate throughout the nation surrounded the organization of grids to support competitive electricity markets (which, outside Texas, led to the adoption of FERC's Order No. 2000).[20] The idea of an independent system operator was generally supported, but there were arguments about whether ISOs should be for-profit or nonprofit entities, what responsibilities they should have (for example, whether they should act just as grid operators or also as power exchanges), how they should be governed, whether industry incumbents would have too much influence on their governance, and whether they should be in charge of a single control area merging control areas of all utilities or simply coordinate among utilities.

Provisions of the Bill. These and similar issues were debated with reference to ERCOT during the 1999 restructuring efforts, as well. In the end, SB 7 included a number of provisions to address them:

- It codified ERCOT's role as a nonprofit ISO with a stakeholder board representing all market participants. (Legislative changes in 2003 changed the board's composition to include additional independent members and put ERCOT under direct PUCT oversight.)
- It codified the simple, but effective, postage-stamp pricing method for transmission service, which would be based on load-ratio share

of each transmission utility. This method was adopted in 1997 by the PUCT.

- It made significant changes to provisions dealing with ERCOT's responsibility to serve as a catalyst of an efficient marketplace:
  - Following enactment of SB 7, ERCOT collapsed its historic, multiple, utility-specific control areas into a single ERCOT-wide control area. All utilities (investor-owned as well as municipally or cooperatively owned) had to turn over operational rights to their grid assets to ERCOT.
  - New systems were developed to handle wholesale financial settlements and a centralized database related to retail customer switching (an innovation recommended by British restructurers and strongly supported by the PUCT).[21]
- It established a system benefit fund, which was financed by all retail customers through a charge collected by the wires companies, to provide discounts to low-income customers qualified by the Texas Department of Health. The fund would also pay for a four-year education program, developed and implemented by the PUCT, to help customers make informed decisions when choosing a retail electricity provider (REP).[22]
- It repealed the integrated resources planning process established in 1995 on the grounds that it was unnecessary in a competitive market in which end-use customers chose resources directly—a repeal generally well-received by the market participants.
- It established a renewable portfolio standard and required an energy efficiency program to be administered by the wires companies.

These last-mentioned renewables and efficiency provisions had their origins in the IRP requirement that customers be made a part of a utility's resource procurements. In a novel process called "deliberative polling," first used in Corpus Christi by Central Power & Light, randomly selected customers were polled about their power preferences after two to three days of intense educational workshops. In the first poll, as well as in the eight that followed around the state, customers consistently indicated a willingness to pay more to support energy efficiency and renewable energy.[23]

This message resonated in the regulatory and political forums just as the legislature returned to Austin in early 1999. SB 7 called for 2,000 megawatts of additional renewable generation capacity to be built by 2009, and the PUCT supported this mandate by developing a renewable energy credit (REC) market immediately prior to market opening. All retailers (competitive and incumbent) had to acquire, and in three years retire, RECs based on their share of statewide retail electricity sales. This requirement created demand for renewable electricity and helped Texas reach the 2,000-megawatt target in 2005, four years earlier than the stipulated date.

Based on this success and on the strong embrace of wind energy development in West Texas, the Texas legislature passed Senate Bill 20 in 2005,[24] which expanded the goal to 5,880 megawatts of renewable generation capacity by 2015, including 500 megawatts of nonwind capacity. The new bill's target for 2025 was 10,000 megawatts of renewable capacity. As of November 2008, Texas was the leading wind energy state in the United States, with 6,297 megawatts of installed capacity; second-place California had 2,493 megawatts.[25] An innovative provision entitled "Competitive Renewable Energy Zones," also adopted in 2005, gave the PUCT new authority to predesignate and preapprove construction of transmission lines to the most favorable wind zones, thus addressing one of the key impediments to wind development. This process was concluded in 2008 with the adoption of a $5 billion expansion program in ERCOT to facilitate the interconnection of a total of 18,000 megawatts of wind generation. In 2007, the legislature also expanded and strengthened the 1999 requirement for an energy efficiency program.[26]

Implementation of SB 7. Implementing all of these provisions of SB 7 (and others we have not discussed) required preparation and testing. Many new systems were needed to manage customer switching and maintain settlement data; new and existing sector professionals needed to experience how the new market would function; and customers needed education to become familiar with retail competition. Accordingly, a six-month pilot program designed to test participants' readiness for, and customers' level of interest in, retail competition was open to customers in the IOU service territories and scheduled to start on June 1, 2001. A high level of interest in the program when the enrollment began in February 2001, especially by nonresidential customers, compelled most IOUs to resort to lotteries to choose the 5 percent of their customers who would participate. The start of the program was delayed twice, however, by computer system problems experienced by the ERCOT ISO, which was (and still is) in charge of recording customer switches from existing utilities to new retail providers. The pilot program eventually started at the end of July and proved quite valuable, allowing for the identification and resolution of problems before the ERCOT-wide market opening date of January 1, 2002.

During the pilot program, no new retail providers offered service in areas outside ERCOT. SB 7 required the PUCT to declare a given wholesale market a "qualifying power region" before allowing retail competition. The PUCT was not able to declare any non-ERCOT areas as such, and the slow progress toward the development of wholesale power markets in the non-ERCOT, FERC-regulated regions of Texas delayed the introduction of retail competition in those areas. As of 2007, customers of Southwestern Public Service (an Xcel affiliate), Southwestern Electric Power Co. (an AEP affiliate), El Paso Electric, and Entergy-Texas did not have competitive retail choice, and Texas policymakers, who understood from the ERCOT experience in the late 1990s that a robust, competitive wholesale market was a necessary precondition for retail competition, were not pressing to expand competition to these areas.

Even more important to the success of SB 7 implementation than the pilot program was the collaborative and inclusive process all stakeholders adopted. In the thirty-month implementation of this comprehensive bill prior to the 2002 market opening, timely decision-making was facilitated by a well-structured, multiparty stakeholder process administered at ERCOT and overseen by the PUCT and its staff. Much as the legislature did with a mix of broad policy goals and specific prescriptions to the PUCT, the PUCT asked that SB 7 implementation details be worked out at the stakeholder level with only the difficult technical or policy issues being taken to the commission. This approach allowed for swift issue resolution, widespread program buy-in, and pragmatic solutions to novel business and customer issues.

In sum, as we have tried to describe above, SB 7 was the result of detailed deliberations over three legislative sessions. As such, it was well vetted by all major stakeholders, and it represented as balanced a consensus of various interests and goals as could be expected of such a comprehensive restructuring bill. The multiparty stakeholder process, under the guidance and leadership of ERCOT and PUCT, and the pilot program were extremely critical in getting the retail competition off to a good start.

Issues Moving Forward

After more than six years of an open retail marketplace, it is fair to say that the competitive electricity market in the ERCOT region of Texas (with 85 percent of the state's customer load) has worked well. Large commercial and industrial customers have shown great interest in competition since the beginning, and their switching rates have proved its viability. Following a slow start, even smaller customers have increased their participation in the competitive retail market. As of early 2008, 67 percent of residential customers had switched either to new providers or to competitive rate offerings of incumbent providers.[27] Today, across the state, many retailers offer a multitude of differentiated products, including renewable energy, varying contract length, and links to nonpower affinity programs.[28]

None of the major problems that were feared by many after the California crisis has occurred. This is not to say that there have been no difficulties at all. Issues have arisen with the ERCOT registration and customer switching system, delayed wholesale financial settlements and retail billing, withdrawal of potential major retailers such as Shell and Enron/NewPower, gaming in congested zones, hockey-stick bidding, intrazonal congestion allocation, ERCOT board governance, and providers of last resort. These have, however, been methodically addressed by the robust stakeholder working groups and, where necessary, the PUCT.[29]

Looking forward, ERCOT's move to a nodal, or geographically sensitive, wholesale market will bring the Texas market into line with successful wholesale market structures elsewhere in the United States; but there are concerns that this change will disrupt the successful retail market operations. Disaggregated wholesale pricing may make it challenging for mass market retailers to offer a regionally uniform retail price. Uncertainty about the effect of wholesale pricing is also influencing investment decisions by generators, especially in the western regions of ERCOT, which already have seen significant congestion from new wind generation. The move to a nodal market has placed an even greater focus on expanding the state's transmission grid.

One criticism of the Texas competitive market has been that electricity prices have increased for residential customers. But such increases have occurred across the nation, due primarily to the marked rise in the price of natural gas, which, as noted earlier, is the dominant power-generation fuel in Texas. On average, close to 50 percent of generation in ERCOT comes from gas-fired plants.[30] More to the point, natural gas is the marginal fuel in Texas, as it is in many other markets. Accordingly, the increase in its price has caused electricity prices to rise—an effect that was particularly acute immediately following the fall 2005 hurricane season, when the spot prices reached $14/Mcf (as compared to the $2–$3/Mcf range seen in 1999, when the restructuring law was enacted). But market pressures pushed retail prices down in 2008 as natural gas prices fell back down to the $6–$7/Mcf range.

At the same time fuel prices have been going up, however, power plant efficiencies have been increasing significantly. One noteworthy consequence of the early introduction of wholesale power competition in the mid-1990s, and the associated streamlining of the new plant-permitting process, was the significant investment in natural gas-fired generation in Texas. Throughout the 1990s and early 2000s, more than 200,000 MW of new gas-fired capacity were built in the United States, about 30,000 MW of which were in Texas.[31] Advances in turbine efficiency enabled the new power plants, by the effect of market forces, to shut down (or at least mothball) the older, less efficient, more polluting gas plants. In traditionally regulated states, these plants would have remained in the rate base, and the higher fuel costs passed through to customers, but today, as a result of the newer technology and competition, residential prices in Texas are returning to the levels seen in the final years of regulation.[32] This decrease is occurring even though the price of natural gas has tripled.

Looking into the future of electricity industry in the nation as a whole, one needs to consider, in conjunction with the price of natural gas, the potential shape of the U.S. carbon emissions policy. Only a few years ago, higher natural gas prices encouraged the construction of new coal-fired capacity; state regulators, concerned by the high cost of natural gas and electricity, overcame their environmental hesitation and approved coal projects.

But the tide seems to be turning again: Intensifying debate about climate change has increased opposition to coal. In the short term, interest in natural gas is surging, with import capacity for new liquefied natural gas, or LNG, and increased domestic production of unconventional natural gas resources helping to keep the price of natural gas at about $6–$7/Mcf, which is viewed as bearable for most generators and customers.

But will the price of natural gas remain at this level if more gas-fired generation is built in the next few years? Unless opposition to new LNG import terminals (other than in Texas and Louisiana) goes away, and more domestic exploration (especially offshore) is pursued, it is difficult to provide an optimistic answer. These concerns may explain the emerging renaissance of the nuclear industry. Many projects are on the table, and some new nuclear plants may be built and become available by as early as 2015.[33] Other much-discussed options, such as integrated coal gasification combined-cycle (IGCC) or advanced-technology coal plants with carbon capture and sequestration (CCS), remain experimental and expensive. Both CCS and nuclear power are capital-intensive projects; yet the first nuclear project announced in over thirty years is in Texas: NRG Energy's proposed addition of two new units at the South Texas Nuclear Project site.[34] Another first for the nation is a new conventional coal-fueled plant to capture carbon dioxide, proposed by Tenaska.[35] Captured CO_2 would be sequestered in the Permian Basin after being used to help produce more oil. This validation of investor confidence in the Texas competitive market is noteworthy, since it contradicts the conventional wisdom that plant investment would be much easier if its cost could be built into the rate base of a state-regulated utility, and thereby placed on captive customers for twenty-plus years.

Currently, the ERCOT market is mitigating its immediate concerns about resource adequacy with new gas-fired generation. Texas has chosen not to adopt a capacity mechanism, which would have provided additional incentives to build new generation, as have some other organized U.S. markets. Moving forward, a nodal market structure with energy-only pricing and a wholesale price cap of $3,000 per megawatt-hour will be in place. Market forces, transmission grid expansion, and federal incentives will continue to drive investment in renewable energy well beyond the 5,880 MW target set by the state—possibly to 18,000 MW by 2015. But even if the problems associated with connecting all this renewable capacity (mostly wind) are resolved through transmission expansion, the presence of that much variably available power in a market of ERCOT's size is new territory, and many questions will persist. For instance, while the need for peak-hour capacity can, perhaps, be met in part by harnessing the huge solar energy resource in Texas, will state policy support another renewable power player? How will Texas get coal plants with CCS and nuclear facilities built? If gas-fired generation becomes expensive enough, the market will send the right signal; but will it send it soon enough?

In other chapters of this book, forward-looking issues such as the move to a nodal market and resource adequacy are discussed in further detail. We hope that we have provided a useful background on how Texas restructured its electricity market and on a few overarching issues that will continue to attract attention over the next few years. The competitive Texas electricity market has worked well so far; its continued success will depend on future energy and environmental policies, both at the state and federal levels. These policies will, in turn, be influenced by how the market evolves. The experience so far gives confidence that the vigilantly monitored, pragmatic stakeholder process, which has already helped the Texas market overcome many challenges, will continue to solve whatever problems the market and the policy environments create.

3

Evolution of Wholesale Market Design in ERCOT

Eric S. Schubert and Parviz Adib

> *Indeed, it has been said that democracy is the worst form of government except all those other forms that have been tried from time to time.*
>
> Winston Churchill, November 11, 1947[1]

Electricity deregulation in the United States has been controversial, and with few exceptions its progress has been slow and uneven. Problems with the structure and governance of electricity markets have been aggravated in most states by overlapping jurisdictions that often provide conflicting regulation and oversight by the U.S. Congress, the Federal Energy Regulatory Commission (FERC), and multiple state utility commissions.[2]

In contrast, the Electric Reliability Council of Texas (ERCOT) region appears to have avoided the worst complications associated with electricity restructuring elsewhere. ERCOT has had the advantage of following a path of restructuring that involved closely coordinated efforts among ERCOT market participants, the Texas legislature, and the Public Utility Commission of Texas (PUCT). It has also benefited from the status of the PUCT as the sole regulator of the wholesale and retail markets and the transmission grid. Unified regulation has reduced the political risks in regulatory innovation in

---

The authors would like to thank the editors of this book and Felicia Schubert for their helpful and objective commentary on early versions of this chapter. The opinions of the authors expressed in this chapter do not necessarily represent the opinions of BP Energy Company or APX Inc.

ERCOT by allowing for coordinated adjustments across the three major parts of the ERCOT market, not just its wholesale market.

By the summer of 1999, when the Texas legislature passed Senate Bill 7 to restructure and deregulate electricity, it had spent nearly five years painstakingly developing a complete (holistic) restructuring plan that tightly linked the interactions of the three unbundled segments of the ERCOT market—retail choice, wholesale power generation, and the regulated transmission grid.[3] The longevity and robustness of the ERCOT market design—which features the treatment of transmission as a public good, active and vibrant retail choice from stand-alone retailers, deregulated power generation companies, and light-handed mitigation of market power, combined with quick disclosure of resource offers—suggest that such a design is sustainable.[4] While a number of American wholesale markets have operated as successfully as ERCOT's, ERCOT has the only retail electricity market in the United States that compares favorably with successful retail markets worldwide.[5]

As a result of this comprehensive restructuring by the Texas legislature, each policymaking decision on the structure of the ERCOT wholesale market—made by the PUCT or debated at ERCOT stakeholder meetings—needed to be placed in the context of ERCOT's entire restructured electricity market. Even in the realm of wholesale markets, however, no U.S. jurisdiction has had smooth sailing in making the transition to a sustainable market design, as policy trial and error has been the rule rather than the exception. In ERCOT, the move from a zonal to a nodal (locational marginal pricing, or LMP) market design has been long and tortuous. In the wholesale markets over which the Federal Energy Regulatory Commission has had jurisdiction, the efficiency benefits of the nodal over the zonal market design were quickly recognized.[6] Resolving the resource adequacy issue has, however, been extremely difficult in these markets, with no capacity mechanism yet showing itself to be sustainable.[7]

Evolution of Wholesale Market Design in ERCOT

The restructuring of the U.S. electricity industry from a group of regulated monopolies to a truly competitive industry has been far slower and more controversial than its champions in the 1980s and 1990s could have possibly imagined. Outside of Texas no other state has been able or willing to provide meaningful retail choice for residential customers.[8] The development of sustainable wholesale markets within the United States has been hampered by the overlapping regulators at the state and federal levels contending with numerous, interconnected issues, such as transmission construction, resource adequacy, and generation interconnection policies, that require their close coordination.

While ERCOT has offered vibrant retail competition for all customer classes and a single regulator that can easily coordinate generation, retail, resource adequacy, and transmission policies, its transition to a sustainable wholesale market design has not been entirely smooth. This chapter chronicles the slow and often painful evolution of wholesale market design in the ERCOT market as a function of three sociopolitical aspects of regulatory oversight in Texas since 1999: the determination of the proper set of policies for ERCOT by trial and error; the need for Texas regulators to develop a broad consensus to implement dramatic changes to the ERCOT market design; and a strong desire by market participants and regulators to avoid disruptive changes whenever possible.

Determining Policies by Trial and Error. The crucial elements of the electric grid in the ERCOT market that had to be incorporated in choosing which generation units to start and operate on a daily basis (and the necessary market framework to complement them) were not self-evident.[9] As a result, stakeholder discussions on the design elements of the ERCOT wholesale market that began in the summer of 1999 could not reasonably have led to a sustainable market design.[10] While some of the correct elements—such as day-ahead ancillary services and direct assignment of congestion rents to those entities sending power across commercially significant constraints—were installed in the ERCOT market early in the process, others, such as the nodal pricing and day-ahead energy market, were products of trial and error that led to a review of a wide range of alternatives. As stakeholders learned what worked and why, their thinking and understanding evolved, and this process was reinforced by actual experience with the shortcomings of the ERCOT zonal wholesale market. The "law of unintended consequences" also came into play: When partial, incomplete solutions were implemented, a myriad of unanticipated problems arose that needed further patches.

The Need for Consensus among Regulators and Stakeholders. Once a market is operating, changes, especially dramatic changes, produce winners and losers. While some participants and stakeholders will always protest new policies, gaining acceptance of a change in course is possible, though time-consuming. Similarly, different levels of understanding of technical issues, differing outlooks on the value of increasing competition, and the relative costs and benefits of necessary changes all lead to lengthy review by regulatory agencies. The transition to the promised land of fully deregulated and competitive electricity markets has required a review of a wide range of alternatives in a field where even the experts do not have all the right answers. Regulatory agencies in Texas have needed both a strong desire to achieve restructuring and technical expertise within their organizations to enable them to mitigate any harm that might result from implementing a proposed change.

A Desire to Avoid Disruptive Changes. Whether they were involved in changing the scheduling, pricing, and dispatching of resources, raising offer caps and reducing price mitigation, or working through the details of a major market power rulemaking, a sizeable number of regulators and participants in the Texas electricity industry restructuring were uneasy about rocking the boat by making changes to any existing market. The ERCOT wholesale and retail markets appeared to be working well when many of its problems were first identified, which made it difficult to push for change.[11]

In sum, when some main elements of the original ERCOT wholesale market design did not work as envisioned, arguments for change based on grounds of market efficiency and economic theory ran headlong into the commercial and regulatory realities of functioning wholesale and retail markets. How the three sociopolitical aspects of regulation described above influenced the debate on those changes begins with a discussion of the main players in the regulatory arena of ERCOT.

The Main Players

Three distinct groups debated and discussed the choice of design elements for the ERCOT wholesale market: the commissioners at the PUCT, the Market Oversight Division (MOD) of the commission, and market stakeholders.[12]

The PUCT Commissioners. Unlike all other state public utility commissions in the continental United States, the Public Utility Commission of Texas regulates wholesale, retail, and transmission service. This wide regulatory mandate gave the commissioners the final say on all market-design issues in the ERCOT power region.[13] While the Texas legislature could override a PUCT decision, legislators tended to defer to the commission's expertise on controversial and complicated issues, as long as the commission was responsive to their concerns. In the development of the initial wholesale market design in 1999–2001, the commissioners asked the ERCOT stakeholders to develop both the overarching design and the details, with an understanding that their proposal would not be altered significantly unless the design were potentially harmful to the market and the public interest.

The Market Oversight Division. In the second half of 2000, PUCT senior management approved the formation of its Market Oversight Division, which hired Dr. Shmuel Oren to serve as its senior advisor.[14] The first task undertaken by the newly formed division as part of its internal mandate was to identify any major shortcomings in the developing zonal market protocols. Over the following five years MOD was very proactive in providing objective opinions to the commissioners and stakeholders. It also took the lead in recommending improvements to the protocols to increase efficiency and reduce gaming opportunities, and in establishing vital PUCT substantive rules associated with the wholesale market design.[15]

The ERCOT Stakeholders. As requested by the commissioners, ERCOT stakeholders spent a significant amount of time developing the initial zonal market protocols. Having worked very hard to identify major issues in advance and propose ways to address them effectively, the stakeholders naturally took a great deal of pride in their design. Their dominant response to changes proposed by MOD was to resist any that was not addressing a specific, narrow problem.

In the first few years after the institution of retail choice and a real-time wholesale energy market, stakeholders were less concerned about the economics of the market than they were about its commercial realities and resulting financial impacts. Retailers in the new market environment were more focused on making the business segment work. Electric cooperatives, municipally owned utilities, and large industrial customers did not want to move to a centralized pool market structure like those in the Pennsylvania–New Jersey–Maryland Interconnection (PJM) and the New England Power Pool (NEPOOL), and a number of the cooperatives and smaller municipals did not like deregulation, period.[16]

The ERCOT Wholesale Market, 1999–2002

In the first eighteen months after the mid-1999 passage of Senate Bill 7, ERCOT stakeholders developed the original set of market protocols that they submitted to the commission in November 2000. Establishing so quickly the rules and infrastructure of a functional wholesale market that worked well with the retail market simultaneously being built from scratch was a remarkable feat. In time, however, weaknesses would become apparent that would require the creation of a new design.

Which Model to Use—Zonal or Nodal? The stakeholders' choice to start the wholesale market with a zonal design was based, first, on the practicalities of getting the market designed and running under a tight statutory deadline, and, second, on their experience in optimizing generation and transmission resources within an individual control area, with occasional coordination among the ten existing control areas. Each control area could be considered the equivalent of a generation portfolio. ERCOT's experience as a group of regulated electric monopolies could not have been farther from that of the centralized, coordinated energy pools that were long established in PJM and NEPOOL.

At the time the ERCOT wholesale market was being designed, a debate over zonal versus nodal market design in the United States was in full flower. As part of the ERCOT stakeholder debate, the question was critical for MOD and, especially, for Dr. Oren, a leading advocate of the zonal approach. Their focus was on the efficiency, fairness, and sustainability of the zonal market design that ERCOT stakeholders were building.

The debate during this period centered on how much of the transmission network needed to be modeled. As restructured wholesale markets had a very short track record, it was more theoretical than empirical. Still, examples of both models were in operation in the United States. PJM and the New York ISO were operating as nodal markets; the California ISO and NEPOOL were operating as zonal markets.

The zonal market design was associated with a "Min ISO" concept, while the nodal market design was associated with a "Max ISO" concept. The Min ISO approach assumed decisions made by market participants would have only a small impact on grid reliability, while the Max ISO assumed they would have a large impact. The choice between the two approaches was based on whether the transmission grid was geared more to life in West Texas (Min ISO) or midtown Manhattan (Max ISO).

The critical assumptions that underlay the ERCOT zonal market design as it related to the national debate on electricity design comprised also the core assumptions of operating a transmission grid associated with a wholesale electricity market. If these five assumptions were true for ERCOT, the Min ISO (zonal) design would be sustainable. If they were not true, then the wholesale electricity market would have to be based on the Max ISO (nodal) concept. The ERCOT stakeholders assumed that:

- *Local congestion would be random and infrequent.* Compared to the transmission grids in the Mid-Atlantic and New England, the ERCOT grid was robust, with far fewer transmission constraints than its counterparts in the Eastern Interconnection. In addition, the PUCT had been aggressive in approving new transmission projects in preparation for the opening of the ERCOT wholesale and retail markets. ERCOT stakeholders, therefore, could feel justified in assuming that local congestion would be random and infrequent when the real-time energy market opened in July 2001.

- *Zonal prices would be sufficient for siting resources.* In 1999–2000, the overwhelming majority of new resources built in electricity markets were central power stations, such as coal- and gas-fired plants.[17] The activation of these plants required a lead time of two years (gas-fired) to six years (coal-fired), and they had to be built in proximity to sufficient water and fuel transportation, such as gas pipelines and railroads. As a result, viable locations would be limited in number and readily apparent to market participants and transmission planners, with plenty of lead time to develop additional transmission within ERCOT to deliver the power to the market. ERCOT stakeholders concluded that more granular pricing at the subzonal level was unnecessary for efficiently siting new generation in ERCOT.

- *Adjusting the number and location of the transmission lines that were assigned congestion rents (transmission lines known as "commercially significant constraints," or CSCs, in ERCOT) and the number of zones that could have separate prices would be systematic and timely.* ERCOT stakeholders assumed that a handful of lines would be regularly congested, and that their locations would gradually change over time as new resources and load growth changed the physical characteristics of the ERCOT grid. The ERCOT protocols envisioned a process where the changes would be made annually, allowing ERCOT to sell transmission congestion rights (TCRs) to stakeholders to help manage the transmission congestion risk.

- *A decentralized dispatch process in real time would be efficient*, and

- *A decentralized unit commitment would be efficient.* These two assumptions highlight the difference between a physical bilateral (zonal) market, where participants are responsible for almost all scheduling and dispatch of generation resources, and a centralized pool (nodal) market, where this responsibility lies with the independent system operator. When the ERCOT stakeholders chose the zonal market design, they assumed that decentralized scheduling and dispatch of generation resources would still result in an efficient outcome, so that the ERCOT operator would not need to take out-of-market actions to maintain the reliability of the ERCOT grid in real time. The stakeholders thought this assumption was reasonable, based on the history of the limited actions taken among individual control areas in the years prior to the opening of the real-time market.

> The zonal model developed by ERCOT stakeholders in 1999–2000 was based on a bilateral physical market with balanced schedules and on the Min ISO concept. ERCOT stakeholders were comfortable giving qualified scheduling entities (QSEs) the freedom to manage their portfolios of plants in close to real time.[18] Portfolio offers in each zone (consistent with the Min ISO approach) would allow the market participants to optimize their portfolios in line with their business plans, not unlike the multiple-control-area approach used in ERCOT prior to the opening of the ERCOT wholesale market. A small (real-time) energy market was to be operated by ERCOT to balance unexpected changes in supply and demand in real time, with clearing prices seen in the offer stack.

Given that other wholesale markets, both nodal and zonal, had very little operating experience by early 2001, these five underlying assumptions of the ERCOT zonal market seemed reasonable to the vast majority of stakeholders and the commission. Only at the end of 2004, after several years of operating experience and review of a wide universe of potential changes (described below), did MOD and the majority of stakeholders conclude that the reality of operating the ERCOT grid did not support all five assumptions, and that ERCOT needed a nodal market design. In essence, experience showed the stakeholders that the dynamics of the ERCOT wholesale market were far more complicated and interrelated than the dynamics of coordinating neighboring regulated monopolies prior to electricity restructurings. The realization of the weaknesses in the zonal market will be discussed further below.

A Crucial Unresolved Issue in the Zonal Market Design: Congestion Rents. A major design feature that was controversial at the start of the ERCOT zonal market was the direct assignment of congestion rents by the grid operator on the power that flowed across transmission lines. Direct assignment of congestion rents on a transmission line is the electrical equivalent of a series of tollbooths that charge varying tolls based on the level of traffic on every major road in a municipality. The purpose of the varying tolls is to discourage use of the roads during times of heavy use to reduce traffic congestion that results from too many cars trying to use the limited space on the roads with tollbooths. In an electricity market, the idea is to create differing nodal prices for each generator that sends power across a congested transmission line. Generators that help relieve congestion on transmission lines receive congestion rents from the grid operator and thus receive higher prices. Generators that require the grid operator to relieve congestion pay congestion rents and thus receive lower prices.

Under the ERCOT zonal model, the only lines that could experience direct assignment of congestion rents were between zonal boundaries. As a result, generators in different zones could receive different prices, based on whether they created or reduced congestion, but all generators within a zone received the same price. Congestion of local lines was cleared using out-of-market actions by the grid operator, and the cost of the operator's actions was uplifted (that is, passed on) to all end-use customers (load) in ERCOT (described in the ERCOT protocols as uplifted on a load-ratio share basis) instead of being directly assigned to the generators causing the congestion.[19] Under the nodal market design, on the other hand, all lines within the grid could be subject to direct assignment of congestion rents, so in theory each generator could receive a different price for the power it generated, based on the amount of congestion it caused or reduced.

In the negotiations among stakeholders, some electric cooperatives and small, municipally owned utilities won a compromise in which the zonal market started with four zones, but without direct assignment of any congestion rents across commercially significant constraints (that is, the transmission interfaces between zones) or within each zone. Instead, an agreement was reached to allow all congestion costs to be uplifted on the basis of a load-ratio share across ERCOT, despite the experience in other markets suggesting that zonal congestion costs would explode.

In late 2000, after eighteen months of hard and painstaking work, the ERCOT stakeholders submitted the zonal market protocols to the PUCT for review and approval. The commissioners decided to institute a docket (what the PUCT calls a "contested case") and to use MOD as their advisors in the review. According to PUCT rules regarding administrative proceedings, as applicants in a contested case, the ERCOT stakeholders were not in a position to consult informally with either the commissioners (who, according to the rules, served as the administrative law judges) or MOD (the advising staff for the judges) on the details of the wholesale market design. All arguments and discussion about design elements had to be presented in public hearings and workshops in the docket.

This arrangement caused the ERCOT stakeholders to feel shut out of the decision-making process because some of the market design elements they had so painstakingly negotiated among themselves were changed by the PUCT in the final order in the docket. These changes were, for the most part, consistent with good market-design principles, and most ERCOT stakeholders recognized them in time as good for the wholesale market design; but in making them, the PUCT unintentionally created a great deal of frustration and resentment toward MOD on the part of the stakeholders, and these feelings colored the decision-making process at stakeholder meetings for a number of years to follow. In retrospect, a more open process in reviewing and approving the protocols would have been the proper approach.

A Compromise on Assigning Congestion Rents in the ERCOT Zonal Model. MOD argued vigorously in the market protocols docket for direct assignment of congestion rents across all congested lines, but not as a default pricing mechanism for all lines (as occurs in a nodal market design). As Shmuel Oren argued in a report filed in the docket, the potential for gaming the system would exist whenever local congestion costs were socialized; an ERCOT market participant could game the real-time dispatch by scheduling its generators to run at a level that would induce the operator to pay the participant to back down the generation unit.[20] This gaming opportunity was known as the "DEC game" (for decremental energy instructions). Since, as we have seen, local congestion energy costs were borne under the ERCOT market protocols by all load-serving entities on a load-ratio share basis,[21] there was no limit on the amount of money a market participant could receive by playing the DEC game. ERCOT stakeholders strongly opposed this approach, saying it was fundamentally incompatible with the market they had designed.

The PUCT commissioners, torn between relying on their market-design experts on the one hand, and the ERCOT stakeholders—the parties that had drafted the protocols—on the other, decided to split the difference. In the final order in Docket No. 23220, as a compromise between the positions of MOD and the stakeholders on the issue of direct assignment of congestion rents, the commissioners ordered that direct assignment would be assessed among zones six months after uplifted zonal congested costs exceeded $20 million over a rolling twelve-month average.[22] The commissioners made the same compromise on local congestion, stating that if a particular modification to the zonal model to allow direct assignment of congestion rents on local lines were not feasible, then the commission would consider implementing a nodal market design.

The ERCOT wholesale market formally began operating as a single control area with a centralized real-time energy market on July 31, 2001. As the experience in other markets had suggested would happen, zonal congested costs immediately exploded, as a number of market participants used the scheduling opportunity presented to them with no concerns for congestion consequences. The $20 million threshold for zonal congestion was surpassed in three weeks. Local congestion costs exceeded their $20 million threshold by early March 2002, after just seven months. While market participants, MOD, and the commissioners all agreed that zonal congestion needed to be directly assigned as congestion rents, ERCOT stakeholders disagreed vigorously with MOD about the need for direct assignment on local lines, and even about the method for direct assignment. The stakeholders debated and rejected a MOD proposal to insert direct assignment of congestion rents into the zonal market design, with both parties agreeing to bring the issue to the commission, as required by the PUCT's final order in Docket No. 23220.

When the threshold for the uplifting of local congestion costs was crossed, the ERCOT stakeholders and MOD commenced a long and arduous debate on the merits of either fixing the zonal market design or moving to a nodal design. In retrospect, if local congestion had been the only serious lingering problem in the ERCOT zonal market design, the stakeholders would have been right in their profound skepticism about MOD's strong push for direct assignment of local congestion rents. But, as will be discussed further, it became clear over time that the zonal market design was also insufficiently detailed either to efficiently schedule, dispatch, and price the ERCOT grid or to efficiently site new generation.

MOD White Paper: The Case for a Better Market. As part of the project reviewing the issue of direct assignment of congestion rents on local lines, the PUCT asked the ERCOT stakeholders and MOD to file comments for consideration. In September 2002, MOD filed a white paper supporting direct assignment.[23]

The white paper was more than just a laundry list of problems with direct assignment as a solution. As it did in a later white paper reviewing the benefits associated with an energy-only resource adequacy mechanism,[24] MOD made the case for market improvement based on the premise that ERCOT needed sufficient flexibility and transparency to incorporate new technologies and market strategies in a market with flourishing customer choice. MOD argued that the PUCT needed to give the deregulated retail and wholesale electricity markets in ERCOT the chance to transform the way in which electricity was bought and sold based on a "bottom-up" approach. If it failed to allocate scarce transmission resources through a pricing mechanism that addressed local congestion, MOD argued, the existing ERCOT market structure would become increasingly arthritic, especially as the role for location-specific resources, such as wind turbines and distributed generation, increased.

The white paper identified the following serious weaknesses in the operation of the ERCOT zonal model over the previous year, highlighting flaws in some of the assumptions on which the decision to implement a zonal design had been based in the first place:

- *Local congestion was systematic and frequent.* The white paper asserted that the direct assignment of congestion rents on non-CSCs would provide an incentive to improve the scheduling of planned transmission and generation outages with an eye toward economic dispatch, variation in seasonal and daily electricity prices, and the business plans of generators. MOD concluded that assigning local congestion fees would prod generators, transmission owners, and the commission to minimize the opportunity cost of lost output. As a result, the efficiency of operating the ERCOT grid would increase, as was seen after the direct assignment of congestion rents across CSCs starting on February 15, 2002.

- *Siting of new generation, especially intermittent renewable resources, was poor.* MOD cited standard economic theory stating that the most efficient way to allocate scarce resources, such as transmission capacity, is to use marginal pricing. Locational pricing would discourage the piling of generation at transmission-constrained sites and likely would have prevented the siting of 1,000 megawatts of wind farms behind a 400-megawatt transmission constraint in the McCamey, Texas, area.[25] The federal production tax credit provided strong incentives to site wind farms in the windiest areas of the country, and the McCamey area had the highest capacity factors for wind readily accessible in ERCOT. Assigning local congestion fees would have directed developers of wind power to choose one of the other numerous potential sites within ERCOT to produce renewable energy where transmission resources were more plentiful. The rapid deployment of new wind resources power in West Texas—and the resulting transmission bottleneck that was so severe it even damaged equipment—exemplified the problems that could occur in a deregulated market with poor locational price signals.

- *Uneconomic and confusing dispatch resulted from the out-of-market solution of local congestion.* The location of line congestion changed over time as supply and demand conditions changed within ERCOT. In the new world of competitive wholesale and retail markets, MOD argued in the white paper, supply and demand would change more frequently and more quickly than it did prior to electricity deregulation. As the result of many market-based transactions that could not even have been imagined in the past, transmission constraints would occur more rapidly in locations that would be less predictable than before.

  Assignment of local congestion charges would make the cost of delivering power from various generating units in ERCOT more transparent to market participants, allowing the market to price these risks properly in bilateral energy contracts. By directly assigning congestion rents on all lines to the generators that were overloading those lines rather than socializing the costs of clearing the congestion to load, MOD asserted, the ERCOT market design would prompt generators and their customers to become

> more innovative in scheduling their generation units in real time to take into account transmission outages and overloads. In the long run, this increased innovation and efficiency would lower the costs of real-time dispatch.

In its white paper, MOD presented a strong case that the PUCT and ERCOT stakeholders needed to fix the key flaw in the zonal market design that MOD had identified prior to the opening of the market. Trying to implement what would appear to an economist as a straightforward fix led to a long policy debate at the PUCT because of the three sociopolitical considerations of regulatory oversight described earlier in this chapter: determining the proper set of policies by trial and error; the need for Texas regulators to develop a broad consensus to implement dramatic changes that affect a large number of constituents; and a strong desire of market participants and regulators to avoid disruptive change whenever possible.

Resource-Specific Offer Curves and Nodal Pricing of Resources: Texas Nodal Rulemaking, 2002–3

Rather than address the single issue of direct assignment of congestion rents on local lines, the PUCT commissioners decided to look for a comprehensive design solution for the ERCOT wholesale market. They held workshops from November 2002 through January 2003 to discuss the scope of the changes, with the goal of providing a framework that included critical market elements. In the workshops, ERCOT staff and stakeholders discussed the need to improve real-time dispatch in the zonal framework. Some stakeholders, in an attempt to provide alternatives to a nodal design, developed a number of proposals with resource-specific offer curves without direct assignment on all lines.[26] These proposals addressed only the engineering problems associated with portfolio dispatch and attempted to give ERCOT operators better tools that they had requested to manage real-time dispatch. Unfortunately, these partial solutions created potential inconsistencies between dispatch (resource-specific) and settlement prices (zonal). At the same time, MOD proposed a market-based solution that dealt with some of the market deficiencies but did not address the underlying engineering problems associated with portfolio offer curves and did not give participants the means to hedge against congestion risk.

In September 2003, the PUCT rejected these partial solutions and approved Substantive Rule § 25.501 to require the implementation of nodal market design by October 2006, consistent with the final order of the zonal protocols docket.[27] The new rule created a basic framework for a nodal market design, with ERCOT stakeholders to develop market protocols consistent with the rule. ERCOT was also ordered to conduct a cost-benefit study of implementing a nodal design in comparison with other viable alternatives (which could include patches to the existing zonal-model software) to confirm that the decision to move to a nodal design was in the public interest. MOD was to participate actively in the stakeholder meetings that developed the nodal market protocols.

The Decision-Making Process and the Three Groups, 2003–5

The choices debated and made in the development and approval of the nodal market protocols again were the product of the interaction of the PUCT commissioners, MOD, and the ERCOT stakeholders. Each group had its own interests and perspectives during this period.

The PUCT Commissioners. In the period when the fundamental issues of the nodal market were being reviewed and questioned, the commissioners had to deal with unhappy stakeholders who either would lose out from the changes proposed or would have to spend additional money to adapt to the market's new features. From the commissioners' point of view, it was not clear whether the benefits associated with the changes MOD was proposing could be demonstrated.

Maintaining an open and deliberative process became increasingly important for the commissioners as they came to understand the scope and cost of a shift from a zonal to a nodal market design, even if the pace slowed the transition. In addition, some of the commissioners were naturally cautious about, or ambivalent toward, making dramatic changes to the wholesale market design, which had been operational for several years. As they held the final responsibility in approving changes to the wholesale market design, the commissioners chose to vet all other alternatives before moving to the proven nodal approach because nodal was highly controversial among most ERCOT stakeholders.

The Market Oversight Division. MOD's major concerns were to ensure that the wholesale market design, first, avoided gaming opportunities and inefficiencies that could develop as part of the compromises developed in the ERCOT stakeholder process and, second, embraced anticipated innovations in electricity markets, such as market-based demand-side response. MOD believed that best practices elsewhere should be the default for the ERCOT wholesale market design, and that reinventing the wheel or developing something novel with hidden flaws was poor public policy when a tried and true method was available.

MOD based this approach on the experiences of a number of wholesale markets in the United States and the world, noting that stakeholder compromises often led to faulty, unsustainable market designs.[28] The division strongly believed that fundamental market design issues needed to be reviewed by outside electricity economists, although it deferred to ERCOT stakeholders and staff on issues associated with the day-to-day commercial and reliability factors of the wholesale market.

MOD, in its white papers in both nodal and resource adequacy debates, framed the case for change by highlighting the tendency of greater price transparency and stronger market signals to increase the risk and rewards for market participants. This "bottom-up" view of the market relied on historical precedents in non-electricity markets or non-U.S. electricity markets that allowed for the proposal of fresh perspectives compatible with the basic structure of unbundled utilities and an active retail market. This viewpoint was in stark contrast to the commercial, or "status quo," approaches taken by almost all of the ERCOT stakeholders.

ERCOT Stakeholders. From the point of view of the ERCOT stakeholders, a change to a nodal design, rather than a few patches to the zonal model, could be costly, time-consuming, and disruptive to their businesses. Very few supported the PUCT's rule to implement a nodal model by October 2006,[29] and many suggested that long-term bilateral contracting was being inhibited by uncertainties in the nature and timing of changes associated with the drafting of the nodal protocols and the cost-benefit study. In short, the stakeholders felt that the market was still evolving, and that MOD and the commissioners should have given the zonal market design more time to prove itself. ERCOT stakeholders also felt strongly that they should develop the details of the market protocols with minimal intervention from MOD.

The views of electric cooperatives, municipally owned utilities, and large industrial customers on the development of a nodal market design for ERCOT were colored by some negative assessments of the PJM nodal market. These stakeholders expressed their dismay about PJM's nodal design early and often, citing weaknesses involving transmission congestion and construction, governance, the tax implications for certain electric cooperatives, and the capacity approach to resource adequacy called "installed-capacity" (ICAP) or "capacity markets."

The opposition of large industrial customers to a nodal market design was centered on their strong dislike of capacity markets. Industrial customers in particular considered ICAP and nodal as intertwined elements of a market design, and therefore were concerned that a nodal market design would destroy the bilateral market in ERCOT. They believed that ICAP markets were a means to partially re-regulate the ERCOT market by forcing loads to provide generators with fixed payment streams outside the bilateral energy and ancillary services markets, which in turn would undermine the benefits received by industrial loads from the successful competitive retail market.[30]

A nodal market design for ERCOT did not inevitably lead to a capacity-based resource adequacy mechanism, but since there was no working model in the United States that featured this combination, such skepticism was understandable.[31] From an economic perspective, the notion of capacity payments as a mechanism for cost recovery and supply adequacy assurance is an anomaly, unique to the electric power industry and originating in its regulated-monopoly legacy. In any other industry, as capital-intensive as it might be, suppliers assume investment risk and recover their cost and profits by selling the commodity or service at market-based prices, while the customers for the service (for example, load-serving entities, or LSEs) assume price risk. The two sides manage their mutual risk through long-term bilateral contracting between them. This "bottom-up" approach, which treats electricity as a commodity and creates markets without a centralized planning mechanism, has been adopted in Australia, the Canadian province of Alberta, and ERCOT.[32]

In the zonal market design, Texas-based owners of large portfolios had distinct advantages over smaller independent power producers (IPPs) and were reluctant to change, while the IPPs wanted to move to nodal to put them on a level playing field with the large incumbent fleets.[33] The zonal design required all energy schedules submitted to ERCOT to be balanced, with ERCOT operating the balancing (real-time) energy market to adjust for unexpected changes in load and generation. Unlike IPPs who owned combined-cycle units or one or two simple-cycle, gas-fired units, owners of large portfolios had the ability to compensate for the real-time physical limitations of individual generation units.[34] (The majority of combined-cycle units were not operated by owners of large generation portfolios.) In addition, IPPs preferred the centralized day-ahead market of the nodal designs in the United States, as they could more easily commit units, which often were combined-cycle units, on the margin of the market.

Among investor-owned utilities (IOUs), the largest one, TXU, was not comfortable with the short amount of time it was being given to fully implement a nodal market design, and it stated its preference for a longer time. The large retail electric providers (REPs), such as TXU Retail and Direct Energy, felt they could operate under either a zonal or nodal market design; however, smaller REPs, who were struggling to get residential and small commercial customers to switch from the incumbent retail providers, were concerned that moving quickly to a nodal market would weaken their positions relative to the better-capitalized incumbent providers. Customers in the North Zone of Texas were mainly concerned about high price differentials that could be seen under nodal market design in their zone due to significant limitations in the existing transmission system.

Nodal Market Design Revisited: Nineteen More Months of Analysis, Debate, and Discussion

Although the PUCT approved its Substantive Rule § 25.501 in September 2003, in which a decision was made to abandon the current zonal market operation and implement all major parameters of a nodal market design by October 1, 2006,[35] the debate on the details of the nodal market protocols was just starting.

Additional Concerns Raised Regarding Nodal Design. Additional issues were raised on the prudence of making a transition to a new market design whose implementation could cause significant costs and create unnecessary risks for all market participants. In particular, the following concerns were expressed by stakeholders:

- A cost-benefit analysis was needed to determine whether a move to a nodal market design was justified.
- "Non-opt-in entities" (NOIEs)[36] were concerned about having a mandatory day-ahead energy market.
- Potential transmission bottlenecks needed to be identified and addressed.
- Ways were needed to address the unequal effects of potential divergences in prices, particularly in the Dallas–Fort Worth Metroplex.
- There was a possibility that implementation costs to ERCOT and its market participants would skyrocket.

Cost-Benefit Analysis. The most controversial question facing the PUCT was that of performing a cost-benefit analysis to ensure that a transition to nodal market design was justified. ERCOT hired Tabors Caramanis & Associates (later purchased by Charles River Associates, or CRA) to conduct a comprehensive analysis of costs and associated benefits, as well as identify potential market impacts on various groups of stakeholders. Quantitative modeling, using GE Multi Area Power Systems analysis software (GE MAPS), and qualitative assessments of impacts were provided to the stakeholders and commission by CRA and KEMA Consulting. The following five findings were key to the decision-making process:[37]

- The nodal market design would provide an average annual benefit of $76 million in generation cost reduction compared to staying with the zonal market design.
- Annual generators' net revenue would be reduced by $781 million.

- The net costs to loads would be reduced by $823 million per year on average.
- The overall benefit to the market would be $300 million in net present value over the ten-year study period.
- The independent power producers, who did not represent loads, were projected to lose about $304 million per year in their revenues while all other market segments were projected to receive benefits by going to nodal market design.[38]

While the results of the cost-benefit analysis were controversial, it provided valuable insights and was a factor in the commission's final decision on market design.

NOIEs' Preference for a Voluntary Day-Ahead Energy Market. Cooperatives and smaller municipalities, such as the cities of Denton and Garland, objected to mandatory participation in a day-ahead energy market (DAM). This was mainly due to the price differences some cooperatives experienced between the day-ahead and real-time markets in the nodal markets of the northeastern United States. The NOIEs insisted on the ability to avoid a DAM and settle their energy and congestion revenue rights (CRRs) in a real-time market (RTM). The issue, which prompted much discussion and debate, caused stakeholders to take extra steps to customize the Texas nodal model to maintain a voluntary DAM—the only nodal market in the United States to do so.

Potential Transmission Bottlenecks. Under the historical regulatory regime, utilities were concerned only with maintaining the reliability of the system within their own control areas at a reasonable cost, rather than spending money to facilitate market operation. It was not surprising, therefore, that significant transmission bottlenecks developed when ERCOT began to operate as a single control area. In particular, such bottlenecks were identified in Texas's South Zone (Rio Grande Valley) and North Zone (Dallas–Fort Worth Metroplex). Without additional transmission lines and upgrades, a nodal market design presented a risk that significant price differentials would occur. Residents of the Rio Grande Valley and the Dallas–Fort Worth Metroplex were concerned that the congestion rents and nodal prices associated with these transmission bottlenecks would raise their electricity costs after the nodal design was implemented.

Inequitable Impacts Due to Potential Divergence in Prices. The existing transmission bottlenecks and the need for a methodology to determine potential load zones were among the most important concerns expressed by some stakeholders, who claimed that transition to a nodal market design would result in a transfer of wealth from high-priced load zones to other areas of ERCOT. In particular, these stakeholders believed that because the existing transmission system was the product of a regulatory regime that had been in place for more than two decades, it would be unfair to address the cost differences among the various geographical areas by imposing them solely on those where bottlenecks occurred. Rather, they thought the commission should take extra steps to accelerate transmission upgrades to create a level playing field, while requiring all customers to continue paying for such upgrades through the postage-stamp transmission mechanism in effect in the ERCOT wholesale electricity market.

Implementation Costs to ERCOT. Finally, stakeholders were familiar with cost overruns that had been announced for the markets run by independent system operators in New England and California, who were making similar transitions from zonal to nodal market designs. Some argued that it would not be worthwhile to initiate such efforts. The result, in their opinion, would be skyrocketing expenses that would eventually be uplifted to all market participants. At the time of these discussions, implementation of a nodal market design was expected to cost ERCOT between $60 million and $80 million. The estimate in 2006 showed a total cost of $263 million. The latest estimate in December 2008 indicated that it would cost ERCOT about $660 million to implement fully a nodal market design.

Critical Evaluation of Stakeholders' Proposed Nodal Market by Four Economists. The first efforts to implement a zonal market design in ERCOT on July 31, 2001, produced several important lessons. In the authors' opinion, the most important bore on the lack of coordination between economic principles and the practicalities associated with engineering the operation of an electricity system. Some of the economic considerations included concerns regarding the real-time market (such as the almost complete absence of supply and demand responses), the day-ahead market (such as the unit-commitment process and local reliability issues), and the long-term market (such as the entry and exit of resources, and difficulties in ensuring a desired mix of resources). Engineering considerations included the lack of electricity storage, limited transportation, the risk of blackout or cascade failure, nonconvexity of generation schedules due to the nature of production cost (start-up time, minimum loading of units, minimum run times, and ramp rates), and unexpected forced outages. While stakeholders were well equipped to handle the technical and commercial operation of the power system, they failed to rely sufficiently on economic principles to obtain incentive-compatible results. In July 2003, MOD submitted a letter to the board of directors of ERCOT insisting that market participants needed an independent economist to advise them in designing nodal market principles. The board endorsed that recommendation, and ERCOT hired Professor Frank Wolak of Stanford University to be available to stakeholders as their market advisor.

In addition, MOD provided an opportunity to its two advisors, Professor Shmuel Oren of the University of California, Berkeley, and Dr. David Patton of Potomac Economics to review and provide constructive comments on more than twenty-five white papers developed by stakeholders regarding ERCOT-proposed nodal market principles. Finally, in response to a request by Reliant Resources, one of the major stakeholders, a paid consultant for Reliant by the name of Dr. Roy Shankar was given an opportunity to provide additional comments on all white papers. The four economists then attended a workshop to discuss and debate market design issues. In stark contrast to the usual disagreements and caveats that enliven typical economist jokes, all agreed for the most part on the major shortcomings identified in the white papers.[39] Among these shortcomings were

- Inadequate market power mitigation measures and inconsistency in mitigation mechanisms between day-ahead and real-time markets;

- an inability to reflect decisions of the process of reliability unit commitment used in locational marginal prices of the day-ahead market.
- the possibility that revenue would be inadequate to fully fund congestion revenue rights;
- inadequate mechanisms to allow for the dispatch and pricing of less flexible gas units;
- inadequate attention to flowgate congestion revenue rights;
- inadequate pricing during shortage conditions and the absence of a meaningful scarcity pricing mechanism;
- inadequate mechanisms to discourage price-chasing by generators;
- inelastic requirements for the different types of ancillary services; and
- a lack of co-optimization of energy and ancillary services in the day-ahead market.

Stakeholders responded positively to most of these comments, reconciling disagreements among themselves and revising the market design before submitting it to the commission in September 2004 for final approval.

Filing of the Proposed ERCOT Nodal Protocols in September 2005. In the fall of 2005, stakeholders submitted their proposed nodal protocols and supporting documents, including the cost-benefit analysis prepared by CRA and KEMA.[40] In addition to being in compliance with the PUCT Substantive Rule § 25.501, the proposed protocols addressed most of the shortcomings identified by the four economists. The major features of the protocols included

- the continuation of bilateral energy and ancillary services markets and allowance for self-arrangement;
- the continuation of the ERCOT-operated market for real-time energy and ancillary services;

- a voluntary day-ahead energy and ancillary services market with bid-based reliability unit commitment (RUC) and co-optimization of energy and ancillary services;
- nodal pricing for congestion management and point-to-point and flowgate CRRs;
- zonal prices for load zones;
- resource-specific bid curves and nodal prices for resources; and
- the continuation of an offer cap, at $1,000 per MWh.

While these represented significant improvements to the protocols proposed in the white papers published in mid-2004, MOD and its consultant, Potomac Economics, found deficiencies remaining in the following three fundamental areas:

- *The real-time market,* which lacked real-time co-optimization of energy and ancillary services similar to those being used or proposed in other U.S. electricity markets. The real-time market was also criticized for inefficient pricing during shortage conditions and a lack of appropriate dispatch and pricing for less flexible units, such as combined-cycle units. The proposed two-stage mitigation of market power was discriminatory in that it would unnecessarily impose offer caps on those players who did not have market power.
- *The day-ahead market*, which lacked attention to local reliability requirements. ERCOT would need to procure adequate resources in the day-ahead market to address such requirements.
- The reliability unit commitment, in which "clawback" provisions in the process discriminated by imposing excessive penalties on market participants who were short of generation to meet the electricity needs of their customers on the day before the operating day when the RUC would be executed.[41]

Dr. David Patton filed direct testimony on behalf of MOD to discuss each of these flaws and made the following recommendations:[42]

- Energy offer caps in the day-ahead market should be removed.
- The local reliability reserve requirements should be reflected in the day-ahead market ancillary service procurements.
- The reliability unit commitment charges should be capped below what was proposed by stakeholders.
- The RUC-related "clawback" provision should be eliminated.
- Real-time co-optimization of energy and ancillary services should be added to the protocols.
- The pricing provisions should be developed to allow energy and reserve prices to reflect the costs of reliability actions taken by operators during shortage conditions.
- Market software and supplemental processes should be developed to provide for the dispatch of gas turbines and the determination of efficient nodal energy prices when these gas units were either dispatched manually by the operators or dispatched out of sequence.
- Replacing the proposed Texas two-step market power mitigation process with the "conduct-impact" mitigation framework applied in other nodal electricity markets.

The PUCT's Final Order. The proposed nodal protocols emerged from many months of hard work, consensus-building among stakeholders, and give-and-take to establish a balance among various conflicting interests. The commissioners were aware of this very delicate balance and determined not to deviate from stakeholders' recommendations unless the flaws could be demonstrated to be harmful to the market. In addition, the commission heeded strong statements by stakeholders that any significant deviation from the proposed consensus would result in equity issues among them, appreciably delaying the expected nodal implementation date and imposing unnecessary costs on ERCOT and all market participants. Stakeholders argued that both outcomes were undesirable at that critical time when ERCOT was making the transition to the new design.

In the unusually brief final order, the commissioners relied on stakeholders' concerns and rejected all but one recommendation made by

Dr. Patton in his testimony, on the grounds that such improvements could be discussed and implemented in the future, if needed.[43] In addition, the commission discussed two remaining major concerns. First, how would potential price differentials among load delivery points and load zones be addressed? And, second, how would unintended consequences of the transition be mitigated? Four actions were taken to address these concerns:

The commission increased the implementation time. In approving Substantive Rule § 25.501 in September 2003, the commission set October 1, 2006, as the deadline for full nodal implementation. Given the concerns raised by smaller market participants, the commission now agreed to a much longer timeline for full implementation, with a target date of January 1, 2009. This deadline was subsequently postponed when it became apparent that the implementation of the nodal market was significantly behind schedule. As of this writing, ERCOT management expects the nodal market design to be fully implemented by December 2010, with the final timeline still to be reviewed and finalized by the PUCT.

The commission prescribed additional transmission expansion to address bottlenecks and minimize price divergence. The PUCT is known for its aggressive support for transmission planning and for timely regulatory approval of proposed transmission lines and necessary upgrades. In fact, transmission expansion is generally considered a more economical solution than investing in new power plants to meet growing demand for electricity, also providing load in any given area with more competition among existing generation assets. About $2 billion had been invested in transmission expansion in the first five years of the ERCOT competitive market operation, and more than $3 billion was expected to be spent on transmission lines in the following few years.[44] Transmission bottlenecks would be identified by ERCOT regional planning groups and highlighted in annual ERCOT transmission reports, with the PUCT acting to approve them by finalizing the transmission companies' applications for certificates of convenience and necessity (CCNs) in a timely manner and facilitating their construction. In addition, the PUCT's substantive rules provided incentives for transmission companies to build transmission lines by allowing for speedy recovery of their investments. The extension of the timeline to implement the nodal market design to January 1, 2009, and aggressive transmission expansion in the intervening years would reduce the potential for significant price differentials among various load zones.

The commission allowed loads to be settled at load-zone prices. The PUCT had earlier decided to continue settling loads at load-weighted average prices within each load zone. Loads at cheap nodes would opt for generation-node to load-node congestion charges, while large loads at expensive nodes and small loads for which real-time metering was not cost-effective would consequently pay higher zonal imbalance charges. To avoid the "cherry-picking" of large loads at cheap nodes that might lead to ruinously rising prices for small customers, the commission determined that loads should be settled zonally, not nodally.[45]

This approach of settling nodes on zonal prices was similar to that taken in nodal electricity markets operated by the New York ISO and the New England ISO. In addition, the NOIEs would be allowed to request their own load zones if they applied to do so by six months before the final implementation of nodal market design. Currently, the nodal market operation is set to begin with at least four existing load zones (North, South, Houston, and West), plus three NOIE load zones.[46]

The commissioners realized this approach would result in some loss of market efficiency due to less effective price signals sent to electricity users; it was, however, a good compromise during a transition from many years of average pricing under the regulatory regime to a workably competitive electricity market. Over time and with the installation of advanced meters, nodal prices for loads at each delivery point might become a viable option.

The commission required resources to pay the ERCOT nodal implementation cost. The PUCT accepted the argument by consumers and load-serving entities that resource owners, as the main beneficiaries of the transition to a nodal market design, should bear the cost of such implementation.[47] The final order in Docket No. 31540 concluded, therefore, that resource owners or operators must be responsible for the total implementation cost, which was projected to be at least $263 million.[48] As was mentioned above, the latest estimate in December 2008 indicates that it will cost ERCOT about $660 million to implement nodal market design fully.

Conclusion

Although major flaws were identified by MOD and its senior advisor as early as September 2002, by the time implementation of the nodal market design is complete it will have taken more than eight years to carefully review and analyze key issues, assess various alternative market designs, revise rules and protocols, identify market impacts and unintended consequences, and finally implement the software that operates the design. The full implementation, which was scheduled by stakeholders for December 1, 2008 (a month ahead of the PUCT-approved target date of January 1, 2009), has now been tentatively rescheduled for December 2010, with the final schedule still to be reviewed and finalized by the PUCT.

In retrospect, we should not be surprised by the long and winding road that had to be traveled to the promised land of competitive deregulated electricity markets in ERCOT and elsewhere in the United States. The wholesale market is an intricate piece of machinery based on a complex marriage of economic and engineering principles. The engineering principles involve maintaining system reliability by keeping the power system in balance, which requires sending the right amount of energy to the right places every few seconds under transmission network constraints of limited storage of electricity; limited transportation of power in real-time; and nonlinearity of costs in committing generation. The economic principles involve coordinating retailers, independent power producers, and other participants using price signals in a deregulated market that provides incentives for players to use resources most efficiently, avoids gaming and uplift, and has the flexibility to adapt to changing conditions.

To develop and implement wholesale market design elements successfully in the ERCOT power region, the PUCT had to put the basic framework of the market into the form of rules, with the ERCOT stakeholders subsequently working through the details in the form of market protocols. The Market Oversight Division, leaning on commission rules and outside expertise, provided feedback to the commissioners and stakeholders to ensure that the developing protocols created a market that was efficient, fair, and compatible with incentives and retail competition. The commissioners, who were ultimately answerable to the Texas legislature for any problems in the market design, signed off on the final version of the protocols in a contested case.

The work on the nodal market design, though well advanced, is not complete. The nodal design will need to be updated and constantly improved in the future. For instance, it may need to be modestly adapted to accommodate an increased participation of real-time demand response associated with the implementation of advanced metering technology in major load centers in ERCOT.

A market transformation associated with rapid technological innovation in the production, delivery, and use of electricity is projected to happen in the next ten years—a revolution largely unanticipated by ERCOT stakeholders in the 1999–2001 period.[49] Their lack of preparedness is not surprising, as nothing in their experience of working for regulated utilities would have readied them for the speed and depth of the changes that can occur in an evolving market.[50] The choices made by the PUCT in its wholesale market design, however, should allow ERCOT retail and wholesale markets to be highly adaptable and welcoming to this transformation.

The authors believe that the market design created in ERCOT, combined with the best retail market in North America, has provided Texans with an integrated infrastructure well prepared to perform successfully as a competitive electricity market. The rest of the United States and other countries would find the ERCOT market structure an excellent model for crafting their own.

4

Achieving Resource Adequacy in Texas via an Energy-Only Electricity Market

Eric S. Schubert, Shmuel S. Oren, and Parviz Adib

Assuring a reliable supply of electricity in a market-based system has been a central concern in restructured electricity markets throughout the world and the subject of an ongoing debate among academics, industry leaders, and policymakers. The main problem is how to reconcile engineering criteria for reliability and resource adequacy with market mechanisms that will provide price signals for investment, while satisfying regulatory concerns regarding just and reasonable costs for consumers.

Three approaches to ensuring generation adequacy currently exist. The first uses energy-only markets with limited price mitigation (for example, high caps on offers into market) that rely on energy remuneration and scarcity pricing to guide investment. The second uses adequacy mechanisms based on capacity products, which take two forms. One is capacity payments to installed or operational capacity, such as are used in Spain, Italy, Korea, and several Latin American countries. The other is capacity obligations imposed on load-serving entities (LSEs) that can be met in several ways, including bilateral contracting with regulatory verification, as in California; centralized capacity markets, as ISOs in the northeastern United

The aauthors would like to thank the editors of this book for thoughtful suggestions on early versions of this chapter and Felicia Schubert for helping us prepare the manuscript for publication. The opinions expressed in this paper are not necessarily those of APX Inc. or BP Energy Company. The authors were on the team at the Public Utility Commission of Texas (PUCT) that developed the energy-only resource adequacy mechanism in Texas. Adib and Schubert were on staff at the PUCT at the time, while Oren served as senior market advisor to the PUCT.

States; and combinations of bilateral contracting with bulletin-board trading of standardized contracts or a central capacity market. The third approach, which can be viewed as a market-friendly version of traditional integrated resource planning, is based on central resource procurement that can take the form of competitive tendering through either a request for offers (RFO) process or bilateral negotiation, as in France and some other European countries; or strategic reserve contracts between ISOs and critical resources, as in the Nordpool countries.[1]

In this chapter, we examine the question of which of these approaches is the most effective through the prism of the energy-only resource adequacy mechanism of the Electric Reliability Council of Texas (ERCOT). We begin with a review of the intellectual and policy debate concerning resource adequacy, followed by an overview of the political economy background and the evolution of the energy-only approach to resource adequacy in ERCOT.[2] Next is a discussion of how the ERCOT market design has met the conditions that make an energy-only market workable in terms of controlling market power abuse and enabling suppliers to collect legitimate scarcity rents. We also describe some ongoing efforts to reduce the tension between, on the one hand, engineering procedures focused on reliability objectives that tend to mute scarcity price signals and, on the other, the market goals of providing scarcity price signals that are needed to encourage investment. Finally, while we defend the decision made by Texas regulators to take the path of an energy-only electricity market, we acknowledge that a few more years of operation will provide the empirical basis for further evaluation of the effectiveness of Texas's approach to addressing the complex issue of resource adequacy.

The Resource Adequacy Debate

For over a decade, academics, industry leaders, and policymakers have debated whether capacity mechanisms separate from energy markets are needed in restructured electricity markets, whether such capacity markets need to be centralized, and, if they are centralized, how they should be designed. Some argue that, given their technical, political, and social realities, electricity markets need to be supplemented by some capacity mechanism that will ensure generation adequacy.[3] The primary objective is to create sufficient incentives for efficient investment choices. In most cases, however, this goal is interpreted as inducing investment in generation that will meet prescribed reliability criteria based on technical rather than economic considerations. Capacity mechanisms, according to this view, would stabilize generators' income streams suppressed by offer caps that are too low to allow generators to recover their fixed costs, restoring the so-called missing money.[4] Capacity mechanisms are also often viewed as a means of achieving efficient investment. From an engineering perspective, capacity remuneration is a mechanism of choice, since it is a "top-down" approach that supports the setting of capacity targets through centralized integrated resource planning and remuneration of the resources on a cost-accounting basis rather than on a market-value basis.

The need for capacity remuneration in addition to payments for energy and reserve provision in the power industry is often rationalized on the grounds that electricity is a necessity and, hence, commodity prices must be controlled, and supply adequacy ensured, through regulatory intervention.[5] Proponents of the capacity approach also argue that *reliability* of supply, which has public-good characteristics similar to national security or fire protection, is a product distinct from *energy*, and, thus, it needs to be regulated and paid for through capacity remuneration.[6]

On the other hand, from an economic point of view, the notion of capacity payments as a mechanism for cost recovery and supply adequacy assurance is an anomaly, unique to the electric power industry, and it originates in the legacy of that industry as a regulated monopoly. In any other industry, however capital-intensive it may be, suppliers assume investment risk and have the opportunity to recover their costs and make profits by selling the commodity or service at competitive market-based prices, while the customers for the service (for example, LSEs) assume price risk. The two sides manage their mutual risk through long-term bilateral contracting between them. This "bottom-up" approach, which treats electricity as a commodity and creates markets without a centralized planning mechanism, has been adopted in Australia, the Canadian province of Alberta, New Zealand, and ERCOT.

From a theoretical economic perspective, the most important question is whether a competitive energy market without separate capacity remuneration can produce a socially efficient technology mix and total capacity level.[7] The answer to that question is yes, provided that the market is truly competitive so that generators behave as price-takers, and energy prices are allowed to reflect scarcity rents when supply is short. Indeed, it can be shown that when the electricity market is at its optimum in terms of technology mix and total capacity, and the real-time electricity grid is optimally dispatched (transmission constraints notwithstanding), then paying a single clearing price for all the energy produced at each point in the real-time market at the marginal cost of the most expensive unit dispatched will result in a revenue shortfall for all dispatched units. The shortfall is exactly equal to the capacity (fixed) cost of the peaking unit. The framework above assumes some explicit or implicit auction with a single clearing price for all dispatched units where the owners of those units reveal their true marginal costs. It also can enable price-sensitive demand response at the retail level.

The shortfall resulting from marginal-cost pricing based only on generation cost can be recouped, however, without the need for capacity payments by allowing scarcity prices to be set by demand response (at the value of lost load, or VOLL) whenever generation capacity is exhausted.[8] When timely demand response is not technically feasible, it can be approximated by administratively setting the uniform clearing price to an estimated VOLL whenever demand is curtailed due to insufficient supply offers in the cost-based, uniform-price auction. Under such a scheme, the amortized cost of a one-megawatt combustion turbine [CT] per hour equals VOLL per MWh times the loss-of-load probability (LOLP), which is the condition for socially optimal capacity in the system.

Although such a scheme can be implemented even in the absence of active retail demand response, it would obviously benefit from active demand participation, which would provide a market-based VOLL instead of an administrative estimate.[9] The central challenges in implementing such a scheme are, therefore, ensuring a workable level of competition in the market so that generators are not in a position to exert market power on a sustained basis and, second, ensuring that prices will reflect scarcity conditions. Satisfying these two conditions simultaneously is not easy, since market-mitigation schemes used to ensure competitive prices often tend to suppress scarcity prices. Furthermore, shortage conditions are often masked, and the scarcity rents muted by, the system operator's deployment of reserves and by out-of-market actions aimed at maintaining system reliability and avoiding involuntary load curtailments.

The Political Evolution of the ERCOT Market

The resource adequacy question in Texas arises in a policy context that has brought about more competition within the state's electricity market. The major factors in this context are certain landmark decisions made by the state legislature and the corresponding implementation actions taken by the Public Utility Commission of Texas (PUCT) through its policies and substantive rules.

In the 1990s, the Texas legislature passed two major electricity restructuring bills. Senate Bill 373 (SB 373), passed by the seventy-third session of the legislature in 1995, opened the state's wholesale electricity market to competition with the understanding that any existing wholesale contracts would remain intact until the end of their terms and conditions.[10] Senate Bill 7 (SB 7), passed by the seventy-sixth legislative session in 1999, amended the Public Utility Regulatory Act (PURA) to allow retail competition to begin on January 1, 2002, in areas served by investor-owned utilities within the power region of ERCOT. Municipal and co-op utilities could choose to opt into competition.[11]

The PUCT also took several actions with significant consequences during this period. In general, these actions fell into three categories:

- Actions taken after the passage of SB 373 in 1995 resulted in the establishment of rules necessary to create a level playing field for all participants in the wholesale electricity market. These rules provided nondiscriminatory access to the transmission system and defined terms and conditions for interconnection to the transmission grid by new power sources.[12] In brief, easy interconnection of generation encouraged aggressive investment in new transmission and allowed socialized payment by all loads for new transmission. These were the main factors contributing to significant new generation additions in the ERCOT power region or wholesale market.

- Actions taken after the passage of SB 7 in 1999 resulted in the establishment of rules necessary to create a level playing field for all participants in the retail electricity market. In addition, the PUCT finalized market rules to set parameters for the operation of the wholesale electricity market within the ERCOT power region as a single-control-area operation. These included rules to unbundle integrated electric utilities functionally, to address the treatment of stranded investment, to define a code of conduct for regulated utilities, to define customer rights and protections, and to set design parameters and protocols for the wholesale market.
- The PUCT took a firm stand on discouraging the construction of any new power plants by regulated utilities that could ultimately result in rate-base treatment of such additional investments. Such plants might increase the magnitude of stranded investment and discourage competitive suppliers from entering the incumbent utilities' service territories, which would ultimately harm prospects for developing competitive markets and customer choice.

As a result of these actions, competitive electricity markets have been growing within ERCOT, and independent power producers (IPPs) have gained significantly in their share of installed capacity. The ERCOT power region has had more than ten years of wholesale competition, accompanied by more than seven years of retail competition. Retail choice is available to all customers within the service territories of traditional investor-owned utilities, with no regulatory price protection as of January 1, 2007, and no apparent need for such protection. The Texas retail market is routinely ranked among the top competitive retail electricity markets in the world and the best in North America based on a number of factors, including switching rates by retail customers among competing providers.[13]

The increase in the IPPs' share of installed generation capacities within Texas has been a blessing. Between 1995 and April 2009, approximately 43,000 MW of new capacity was added in Texas, of which IPPs and other nonutility entities, such as combined heat and power producers (CHP), accounted for more than 85 percent.[14] Electric cooperatives and municipal utilities, which chose not to be competitive at the retail level, and investor-owned utilities accounted for the rest. In terms of power regions, about 90 percent of the new capacity was built in ERCOT, and the rest was divided between the Southwest Power Pool (SPP) and Southeast Electric Reliability Council (SERC).

How the PUCT Chose the Energy-Only Approach

In 2001, the PUCT began a rulemaking on resource adequacy. As part of the proceeding, the commission's staff and the stakeholders reviewed existing resource adequacy mechanisms, including the installed-capacity (ICAP) markets. Following the lead of other established electricity markets in the United States, the staff proposed a centralized capacity market in 2002. Generation owners strongly favored capacity mechanisms, while retailers and industrial consumers were opposed.

In 2003, with the debate over zonal versus nodal market design dominating the stakeholder process and a very large reserve margin in the energy market, the PUCT decided it could postpone consideration of a resource adequacy mechanism. The commission staff and a number of stakeholders noted that postponing the decision on resource adequacy would allow the PUCT to review the outcome of the Federal Energy Regulatory Commission's (FERC's) standard market design process, which might provide a capacity mechanism that the PUCT could adopt.

Despite having recommended the use of a capacity approach to resource adequacy for ERCOT, the staff had a number of concerns about a resource adequacy mechanism based on capacity payments. First, peaking and baseload units had two very different payment streams, which were not easily reconciled in a capacity mechanism. Second, the locational and operating characteristics of the wholesale electricity market were not easily covered in a capacity mechanism. Third, retailers and industrial customers fiercely opposed capacity markets, believing that capacity payments had not proved effective in adding new generation to other markets. And, fourth, unlike electricity markets on the East Coast, ERCOT by 2003 had already functioned as an energy-only market and had attracted substantial new investment without capacity payments.

During the suspension of the rulemaking on resource adequacy, the commission staff had intended to look to the existing markets in the Eastern Interconnection—as it had with the nodal market design—for a solution to the resource adequacy issue. The flaws in the existing ICAP markets had become increasingly apparent, however, and no proven model of capacity markets was available.[15] Even worse, from the staff's point of view, a review of a draft of the Pennsylvania–New Jersey–Maryland (PJM) reliability pricing model (RPM) showed that, far from developing a market-friendly solution to resource adequacy, PJM was moving in the direction of a centralized integrated resource planning approach.[16]

At about the same time, the PUCT commissioners were still reviewing alternatives to a nodal market design, holding two workshops on the subject in December 2004.[17] Afterward, with doubts still lingering about the nodal design, the staff reviewed two existing zonal designs: the United Kingdom and Australian electricity markets. The United Kingdom design was quickly dismissed as an alternative, as it allowed the grid operator to contract actively for resources to counter the position of market participants when those positions were deemed unfavorable to the market. Such an approach was completely contrary to the one ERCOT stakeholders had taken in designing the ERCOT market, which relied on the grid operator's maintaining system reliability without considering the impact of its actions on clearing prices.

Growth in load and the retirement and mothballing of a number of inefficient gas-fired plants caused the commission staff to restart the resource adequacy rulemaking. In February 2005, the staff held a lengthy conference call with the monitors of the Australian market about the various elements of the Australian market design. While it became evident that the Australian zonal approach would not easily be transferred to the ERCOT market, the staff found in the Australian New Electricity Market (NEMMCO) an energy-only resource adequacy mechanism that had been working successfully for more than six years.[18]

The evidence of a working energy-only market provided the stimulus to the commission staff and the needed reassurance to the commissioners to pursue such a solution. The staff began drafting a white paper later that month to explore practical alternative approaches to addressing resource adequacy effectively. Because the Wholesale Market Oversight (WMO) group (formerly known as the Market Oversight Division) of the PUCT was hosting the semiannual Energy Intermarket Surveillance Group in April 2005,[19] the staff decided to conduct a workshop after the meeting so market monitors from the United States, Canada, and Australia could make presentations on the two competing approaches to resource adequacy: an energy-only market versus separate energy and capacity markets.

In the week prior to the workshop, the staff filed a white paper with the commission,[20] explaining how an energy-only approach might work in ERCOT. It also took the position that an energy-only wholesale market design combined with an active retail market would accelerate adoption of potential innovations, such as market-based demand-side response, as well as upcoming technologies, such as advanced metering and solar power. Expressed in the paper were concerns regarding the need to develop market-based demand response at the retail level, the determination of the offer-cap level, and the question of distinguishing between scarcity pricing and the exercise of market power.

The workshop in April 2005 was the watershed event in the movement away from a capacity approach to resource adequacy that was being considered in other U.S. electricity markets at that time. The Australian regulator made a persuasive presentation, showing that an energy-only market with active retail competition was not only feasible, but had been thriving for more than six years. The Alberta regulator showed an energy-only approach that was being successfully used in North America.

In a subsequent workshop, the commissioners allowed proponents of energy-only markets and capacity markets (who were ERCOT stakeholders) to make a case for each approach. At least one commissioner worried that capacity payments were subsidies that, once established, would be very hard to remove.[21] The capacity market proponents had difficulty coalescing around a single, workable structure, foreshadowing the great difficulties and controversies that would be faced in gaining a broad and stable consensus on a workable capacity mechanism, as was seen in the still controversial development and operation of PJM's RPM and the New England ISO's Forward Capacity Market (FCM). Not surprisingly, owners of large-generation fleets favored a capacity resource-adequacy mechanism, and retailers preferred an energy-only approach. Owners of smaller generation fleets were split on the issue but were concerned that the offer cap developed as part of an energy-only resource-adequacy mechanism would not be high enough to be sustainable.

By the spring and summer of 2005, the commission was faced with a fundamental choice with respect to the evolution of the wholesale market design and the continued success of retail competition: The PUCT could increase reliance on markets (through an energy-only approach), or it could return to the days of integrated resource planning (through the capacity approach).

The energy-only approach would create the potential for higher and more volatile prices during times of scarcity. Retail load aggregators, such as load-serving entities, would need to learn the skills to manage this price risk effectively. Generators, on the other hand, would face more investment risk than any other market in the United States. The symmetrical price and investment risk, if placed into the design properly, would provide strong incentives for innovation and longer-term bilateral contracting.

Within a few months of the April 2005 workshop, the commission chose the energy-only approach and put its staff in charge of drafting the details. Given the strong similarities between the Australian and ERCOT markets and the proven success of the Australian approach, the staff based its resource adequacy rule on the Australian resource-adequacy mechanism, with some deviations reflecting differences between the two markets.[22]

The choices of the energy-only elements for the ERCOT electricity market were also based on the prevailing political realities within ERCOT. ERCOT had a unified regulatory regime in place, where the PUCT was responsible for regulation of the wholesale and retail markets, as well as the transmission grid. In addition, retail competition had been highly successful in ERCOT, without any requirements that competitive retailers demonstrate to either ERCOT or the PUCT that they could meet their projected needs far in advance of the real-time market.[23] The ERCOT market had been relying on price risk for choices in the scheduling and contracting of resources.

Market power abuse, particularly through physical withholding of resources, also was a serious reliability concern because ERCOT, like the Australian market, was a medium-sized, isolated interconnection with a single settlement market,[24] where sizeable system-wide frequency deviations due to loss of power resources were not uncommon. As a result, the energy-only market needed to provide incentives for owners of generation to offer their resources freely into the real-time market to obtain scarcity pricing, rather than withhold resources to obtain scarcity pricing through an administrative pricing mechanism.

The remaining challenge the PUCT faced in developing the recently adopted energy-only resource adequacy mechanism was the incorporation of a complementary market power mitigation regime.[25] Market power and resource adequacy intersect on the vexing issue of scarcity pricing. Failure to address market power results in prices that are too generous for producers based on their exertion of market power, producing price signals that do not truly reflect demand and supply conditions. In the long run, the artificially high prices resulting from the exertion of market power are unsustainable; they undermine economic efficiency, weaken public confidence in the market, and create an uncertain regulatory climate that hinders investment in new generation. Similarly, too much price mitigation results in prices that are too generous for consumers, blocking a price signal reflecting actual demand and supply conditions. Low prices discourage generators from developing new generation to meet growing demand for electricity and from replacing the older, less efficient generation available to run the handful of hours needed each year to meet annual peak demand.

Both of these outcomes result in a lack of adequate investment in merchant plant development, and they can cause shortages in power supply. It is, therefore, important for policymakers to address both issues—market power and resource adequacy—at the same time, to ensure that their interdependencies are adequately addressed. Accordingly, the PUCT combined the ongoing resource-adequacy and market-power rulemakings into a single proceeding.[26]

During the deliberations on the resource-adequacy rulemaking, all three commissioners stated repeatedly that increases in the offer cap had to be accompanied by more rapid disclosure of information affecting the operation of the ERCOT markets. The proponents of increased disclosure argued that it would help ensure heightened market transparency and discourage market-power abuse. Increased transparency would also reassure the public that the price changes they observed were the result of a properly functioning competitive market.

The interrelationship of higher offer caps and reduced mitigation, on the one hand, and the more rapid disclosure of information about resource-specific offers, on the other, was consistent with disclosure policies in other U.S. and foreign markets. In 2006, when the resource-adequacy rule for the ERCOT market was being considered by the PUCT, FERC-jurisdictional markets such as PJM and the New England ISO released resource-specific information six months after they gathered it, which might have been adequate when individual resource offers were heavily mitigated through conduct and impact tests, and offer caps were relatively low. More rapid disclosure of resource-specific information appeared to provide limited benefit under these circumstances in these markets because prices were mitigated in advance of being announced, and those rare circumstances when price spikes occurred were almost always known to market participants.

In contrast, an energy-only resource adequacy mechanism with lighter mitigation of resource-specific offers was less predictable and less transparent. It was therefore argued that more rapid disclosure of resource-specific offers was needed to provide market participants with the same range of information and protection found in FERC-jurisdictional markets. This combination of lighter mitigation and quicker disclosure was already seen in established electricity markets outside of the United States: The Australian electricity market disclosed resource-specific offers with the names of the generators making the offers within twenty-four hours; the New Zealand electricity market disclosed the same information within fourteen days; and the Alberta electricity market displayed the output of each generator, by name, on its website in real-time. Yet it was still debatable whether disclosure of aggregate offer curves would be sufficient to support competition, and whether early disclosure of more detailed information would serve primarily as a means of market mitigation, de facto creating a "shame cap" that might deter the exercise of transitory market power by exposing to scrutiny those firms engaging in it.

Details of the ERCOT Energy-Only Market

The ERCOT energy-only market has a number of features that are unique to U.S. wholesale electricity markets. Among them are offer caps above $1,000 and quick disclosure of offers into the day-ahead ancillary services and real-time energy markets of ERCOT, as well as system-wide market power mitigation divided into two parts: an exemption from system-wide market power mitigation for generation owners whose fleets comprise less than 5 percent of installed capacity, and a voluntary mitigation plan, to be agreed upon by the PUCT staff and the independent market monitor and approved by the PUCT commissioners, that could serve as a safe haven for the owners of the largest generation fleets within ERCOT.

Higher Offer Caps and Scarcity Pricing. One of the PUCT's broad policy objectives in adopting an energy-only resource adequacy mechanism was to provide greater assurance that generation companies and developers would invest in the resources needed to supply the electricity needs of ERCOT customers by allowing prices to rise in response to scarcity of resources in the market. It would, in particular, encourage the development of such alternatives by providing incentives for the development of new peaking capacity.

The PUCT reasoned that a $1,000 per MWh offer cap would provide sufficient incentive for market participants to build and to contract for new baseload, intermediate, and intermittent renewable generation—resources that could meet about 98 percent of the electricity needs of ERCOT. ERCOT stakeholders, PUCT staff, and the commissioners did not believe, however, that a $1,000 offer cap would necessarily provide incentive for the market to support the 2 percent of hours when electricity demand was at its highest: late weekday afternoons in the summer. New peaking generation or demand-side resources might need an opportunity to earn more than $1,000 per MWh (that is, to earn scarcity pricing) in the ERCOT market to cover their capital costs during this very limited number of hours.

Allowing scarcity pricing would provide load-serving entities with the incentive to procure sufficient peaking generation or demand resources as a hedge against scarcity pricing in the ERCOT spot market during this time. Concurrently, ERCOT would need to establish the appropriate credit limits on load-serving entities to limit their ability to lean heavily and consistently on ERCOT spot markets to meet their customers' demand for electricity. Such a strategy would be risky in the face of scarcity pricing, potentially causing a default in payments to ERCOT that other load-serving entities would need to cover. Prudent credit policies also would provide strong incentives for load-serving entities to bring sufficient generation and demand resources to the ERCOT spot market during annual peak demand to maintain reliable operation of the ERCOT grid without the need for out-of-market actions by the ERCOT grid operator or a capacity resource-adequacy mechanism.

Another reason the PUCT chose an offer cap higher than the prevailing $1,000 per MWh was its belief that, under an energy-only resource adequacy mechanism, ERCOT could not rely on a daily "must-offer" requirement or capacity payments to ensure the availability of sufficient resources in those situations. A higher offer cap could provide strong incentive for investment in quick-start generation and load response to meet demand in unusual market situations. Such an incentive turned out to be critical in maintaining reliability in ERCOT, which is a small electrical interconnection compared to the Eastern or Western Interconnections in the United States.

As outlined in the approved rule, on March 1, 2007, the offer cap rose from $1,000 per MWh to $1,500 per MWh.[27] On March 1, 2008, it rose to $2,250 per MWh, and it is scheduled to rise again to $3,000 per MWh two months after the market begins operation under a nodal market design, projected to occur sometime in early 2011. The PUCT chose a significantly lower offer cap than its counterpart in Australia, in part because the ratio of all-time peak to average summer-peak demand in ERCOT was not as high.[28] The PUCT decided to phase in the increase over a three-year period, rather than implementing it immediately, consistent with the three-year time frame in the rulemaking to improve market transparency gradually.

The PUCT also decided that to make the offer caps sustainable, ERCOT needed to increase the price-responsiveness of load in spot markets. The commission stated in other PUCT rulemaking projects that the price elasticity of demand was limited by the lack of interval metering for many loads and plans. The PUCT also considered requiring ERCOT to enable advanced meters for residential and other small loads to provide customers and retailers with more discrete electricity usage information than is provided by monthly billings and average-load profiles.[29] The PUCT has completed Project No. 34610, *Implementation Project Relating to Advanced Metering*, for which the commission staff worked with ERCOT stakeholders to develop a plan to implement the back-office infrastructure, as well as settlement software to allow for fifteen-minute settlement of all competitive loads using advanced meters. At the end of 2008, the PUCT also approved plans proposed by the two largest transmission companies in ERCOT, Oncor and

Centerpoint, to deploy advanced meters to all residential and small commercial customers in their territories within roughly five or six years.[30]

Publication of Resource-Specific Offers into ERCOT-Procured Markets. Effective March 1, 2007, most of the required disaggregated information in the ERCOT market was to be disclosed ninety days after the day for which the information was accumulated—that is, within one-half of the previous disclosure time frame of 180 days. This rule was designed to shorten the disclosure period to sixty days, then to thirty days, on the dates when the offer cap was to be raised from the original $1,000 per MWh to $2,250 per MWh to $3,000 per MWh. The implementation schedule for disclosure was tied to the schedule for increases to the offer cap because, throughout the debate in the rulemaking, all three commissioners emphasized the interrelationships of these two issues. They believed the potential for higher prices would require greater assurances to the public that the prices were the result of a competitive market and not of market manipulation.[31]

One major exception to this disclosure schedule pertained to offer curves by individual resources for balancing energy and ancillary services. Because these two areas raised the greatest concerns about the possibility of market power abuse and other market manipulation, the PUCT stated that expedited disclosure of offer curves for these services was appropriate to provide greater transparency to the public and affected market participants. The initial disclosure provisions in the PUCT rules were contested in court by some participants, eventually leading to a compromise that balanced some of their concerns about disclosure of strategic business information against the greater need for public scrutiny.[32] In its final ruling on the matter, the PUCT decided that, as a general rule, the offer curves should be disclosed sixty days after the day for which the information was accumulated.[33]

As part of the disclosure rules, market price-setters would now be identified after forty-eight hours for each settlement interval. For each period that it ran a balancing energy auction or an ancillary-capacity service auction, ERCOT would publicly identify the supplier with the highest-priced offer accepted, along with the price of the offer. This disclosure would be unremarkable and uninformative most of the time. When prices ran high, however, the public would quickly know whose offer caused the price to clear where it did. A supplier would still be able to price its offer however it wanted (up to the prevailing offer cap), but an offer priced significantly above marginal cost would draw public attention if it ended up setting the market-clearing price. Through the threat of intense public scrutiny of any inappropriate market behavior, this targeted transparency was intended to deter persistent gaming without compromising a supplier's ability to offer energy or capacity at prices sufficient to cover a unit's marginal cost.

Moreover, to enhance market transparency further when significantly high prices were offered, the commission approved an event trigger to be used to identify portions of participants' offer curves that should be disclosed after seven days. The event trigger was defined as a calculated value for each interval that was equal to fifty times the Houston Ship Channel natural gas price index for each operating day, expressed in dollars per megawatt-hour (MWh) or dollars per megawatt per hour (MW/h).[34]

Scarcity Pricing Mechanism (SPM). The SPM, based on the Australian model, was intended to raise offer caps to encourage resource adequacy while preventing excessive transfers of wealth from load to generation during years when reserve margins were thin. The rule relied primarily on high energy offers by generators or by demand resources to set the scarcity prices during shortage periods. While such scarcity prices were justified and necessary for cost recovery in an energy-only framework, time lags in construction of new capacity and limited demand-response capability might result in prolonged periods of high prices and "excessive" recovery. Allowing such excessive recovery would result in an unwarranted transfer of wealth (at least from the PUCT's point of view) to generators from load.

The SPM would operate on an annual resource adequacy cycle, in which the peaker net margin (PNM) would be calculated as the sum of all positive differences between the clearing price in the ERCOT real-time energy market and the estimated marginal cost of operating a generic peaker with a heat rate per MWh of 10 million British thermal units (MMBTU). At the beginning of the annual resource adequacy cycle, the system-wide offer cap would be set at one of the offer caps listed above, which was denoted as "high cap" (HCAP). If the PNM were to exceed $175,000 per MW during an annual resource adequacy cycle, the system-wide offer cap would be reset at a lower level, denoted as "low cap" (LCAP),[35] for the remainder of that cycle. The offer cap would be restored to the highest level allowed by the rule at the beginning of the next cycle.

Exemption on System-Wide Market Power Based on Installed Generation Capacity, or "Small Fish Swim Free." As explained above, the success of an energy-only market hinges on competitive offers that are not inflated through sustained market power abuse, and on scarcity pricing that reflects shortage conditions. This desired outcome traditionally has presented U.S. economists and policymakers with a dilemma. Economic theory suggests that price-taking behavior results in short-run marginal-cost pricing in the real-time market. Short-run marginal costs will not, however, allow for sufficient inframarginal profits to support peaking gas-fired generation that needs to recover a large amount of its fixed costs in the small number of hours it operates in a given year. Unfortunately, those hours when inframarginal profits will be needed are also the times when a number of generation suppliers can exert market power and potentially inflate market prices. To restrict artificially high prices resulting from the exertion of market power (both local and system-wide), U.S. electricity markets have included market mitigation on offers from generators. When offers from *peaking* generation are restricted by *ex ante* mitigation to reflect the short-run competitive outcomes, as has been done in FERC-jurisdictional wholesale markets, the result is inconsistent with scarcity pricing and inframarginal profits. Thus, when exploring the possibility of using an energy-only resource-adequacy mechanism, economists and policymakers confront the "Gordian knot" of having to determine conditions under which scarcity pricing is a function of the exercise of market power or of genuine resource scarcity.

The "bottom-up" alternative that had been used in the Australian market—light or no *ex ante* mitigation of energy offers from generation—has relied on transitory (but not systematic) market power of pivotal suppliers and hockey-stick bidding strategies during shortage conditions, to set scarcity rents through high offer prices. Such an approach had been working in Australia because generation ownership has been sufficiently dispersed among market participants to make the exercise of market power transitory under limited conditions that correspond very closely to conditions of genuine scarcity.[36] Depletion of operating reserves is another common metric for scarcity conditions used by many U.S. independent system operators, but it would not be useful in Australia, which does not have separate markets for operating reserves.

The Australian approach was contrary to the current stance of policymakers in FERC jurisdictional markets and the theoretical frameworks of leading U.S. electricity economists (almost all of whom have favored a "top-down" approach to resource adequacy).[37] As a result, it was subject to controversy or outright dismissal as a viable alternative for needed scarcity pricing in other U.S. electricity markets. The dismissal of the Australian approach may have reflected the fact that a number of preconditions underpinning its success have been far from being implemented in many U.S. electricity markets. These have included vibrant retail competition, reduced concentration of generation ownership, and state and federal policies that encourage the development of adequate generation and transmission resources.

The leading "top-down" alternative for scarcity pricing has involved administrative intervention during shortage conditions that are typically reflected by emergency states and depletion of operating reserves (which could occur during summer peak hours when electricity use is near or at its annual peak). Such an approach was adopted at Midwest ISO, which would rely solely upon an administrative demand function for reserves to calculate an adder to the market-clearing price for energy when operating reserves were being drawn down to meet real-time demand. The "top-down" approach was well-suited at that point to the Midwest ISO wholesale market, given that none of the Midwest ISO states had either retail choice or a retail market nearly as active as ERCOT's.

Sole reliance on the "top-down" administrative pricing approach to produce scarcity pricing was, however, ruled out in ERCOT on the ground that it stood in conflict with the reliability needs of ERCOT's isolated interconnection and its active retail market. It was argued that in the absence of a must-offer provision for contracted resources, market participants would have an incentive to physically withhold generation to trigger scarcity pricing. Unlike in the Eastern or Western Interconnections, physical withholding of generation in ERCOT could cause severe reliability problems that would force the grid operator to intervene administratively to keep the lights on, potentially undermining the market. While a must-offer obligation is present in markets with administrative scarcity pricing, in ERCOT it did not exist due to the retail market's need to avoid a regulatory requirement for bilateral contracting with a "must-offer" provision.

The Gordian knot of genuine scarcity pricing was dealt with by the PUCT staff in the classical way: by making a "decisive cut." The new rule gave small suppliers a safe harbor: If an entity were to control less than 5 percent of the installed capacity in ERCOT, it would be deemed not to have system-wide market power, and therefore would need not worry about prosecution if it decided not to offer any of its capacity into the market. On the other hand, exceeding the threshold would not necessarily mean the entity had market power. It would mean that if the supplier appeared to be withholding production or exercising economic withholding, and prices were being affected, the first question investigators would ask would be whether the entity had market power.

This "small fish swim free" approach was the result of an empirical review of balancing energy data by PUCT staff, which suggested that if the two or three largest generation fleets in ERCOT behaved as price-takers and allowed others to offer as they wished, the market would produce high prices when genuine scarcity resulted, as seen in the other single-settlement, energy-only markets of Alberta, Australia, and New Zealand. Sole reliance on high offers from market participants to set scarcity rents in the Texas market was still being questioned, however, and the debate was by no means over. In a report concerning a reliability event on March 3, 2008, when ERCOT experienced a large sudden drop in wind power, the ERCOT independent market monitor concluded that

> a) relying upon the submission of high-priced offers is an unreliable means of producing scarcity prices during scarcity conditions; and b) the price formation process during shortage conditions can become distorted if it does not include mechanisms to efficiently price the value of sacrificing the reserves that are required to maintain minimum reliability requirements.[38]

The quick disclosure of individual offer curves was another "bottom-up" feature of market power mitigation, in that it was intended to level the playing field and clarify when prices were a product of systematic market power abuse or genuine scarcity—an alternative to the heavy, unit-specific mitigation seen in other U.S. markets.[39] Quick disclosure of individual offer curves ran against the grain of prevailing academic thinking, however, which suggested that revealed information on individual offers tacitly facilitated collusion among suppliers and helped them sustain high prices in excess of competitive levels. The literature concluded that the ability of competitors to cut prices "secretly" so that they were not exposed to retaliatory actions by their rivals was an essential element of competition.[40] These theoretical concerns had not been shown to be problematic in the Australian market, according to Australian market monitors who conducted an internal review of historical price data.[41] After careful consideration and consultation with several market power experts in academia and a number of wholesale power marketers, the PUCT staff finally decided to delay disclosure of various classes of information long enough to minimize any potential consequences of premature disclosure.[42]

ERCOT stakeholders, other market monitors, and even academics continued to express skepticism about this combination of the "small fish swim free" approach to system-wide market power and quick disclosure of resource-specific offer curves for energy and ancillary services, though the first two years of operation showed this "bottom-up" approach to be workable. Further evaluation of the effectiveness of the PUCT's approach would be possible after a few more years of operating experience.

Voluntary Mitigation Plan. In the ERCOT energy-only market, a supplier too large for the small-supplier exemption might also obtain advance protection against prosecution for market power abuse. Presumably, the supplier would submit a voluntary mitigation plan ensuring transparency in the availability of resources (to avoid physical withholding) and make the supplier a "price-taker" during times of scarcity (to avoid economic withholding). This safe harbor, however, would be specific to the supplier's own circumstances and subject to approval by the PUCT. The new rule allowed generators to apply for a voluntary mitigation plan that, if followed, would constitute an absolute defense against a finding of market power abuse with respect to the behaviors addressed in the plan. A large supplier could forgo the voluntary mitigation plan altogether if it believed it had no need for it.[43]

Challenges in the Transition to a Sustainable Energy-Only Approach

When the Australian energy-only resource-adequacy mechanism was introduced, competitive retailers had a contracting requirement for the first three years as a means to ensure grid reliability in the real-time market.[44] The offer cap, set at $A5,000 initially, also provided a very strong incentive for retailers to bring sufficient resources to cover their real-time positions. As such, the Australian energy-only resource adequacy mechanism appears to have worked well from its inception.

ERCOT, on the other hand, faced some challenges in making the transition to a sustainable energy-only approach. First, the energy-only resource-adequacy mechanism was added to the existing physical bilateral zonal market design, which had relied on out-of-market actions that affected real-time pricing in ways not consistent with the energy-only resource adequacy approved by the PUCT. Such actions included deployment of replacement reserves when the operator anticipated the depletion of the energy-balancing stack and deployment of spinning reserves for energy, or deployment of reliability must-run (RMR) resources to relieve intrazonal congestion and violation of voltage constraints. In the context of the newly designed nodal market, the deployment of incremental resources through the reliability unit commitment (RUC) process could be viewed as an out-of-market action.

Second, as mentioned above, competitive retailers in the ERCOT market had never had a contracting requirement that would guarantee sufficient resources being brought to the real-time market. Retailers and other load-serving entities likely experienced a learning curve in finding the correct balance between minimizing their procurement costs and reducing the risk of exposure to the real-time market.

Third, the gradual transition of an offer cap of $1,000 per MWh to $3,000 per MWh—levels far lower than the initial offer cap in the Australian market—had not, in some circumstances, provided sufficient incentive in the real-time market for quick-start, gas-fired generation to be available to respond to reliability events on the grid, and for loads to contract sufficiently to cover their load requirements in real time.

Responding to these circumstances, ERCOT stakeholders and the ERCOT independent market monitor (IMM) expressed concerns in 2007 that some of the defensive actions taken by ERCOT in fulfilling its reliability mission, such as using conservative load forecast and procuring more responsive reserve service, were interfering with market prices and, in particular, with scarcity pricing. ERCOT operations staff acknowledged this problem and worked with stakeholders to address it. The "excessive" out-of-market actions by the system operator were attributed in part to sellers' poor performance of delivering the energy they promised under adverse weather conditions and selling more available capacity than they could provide in the real-time market. ERCOT tried to compensate for the reduced performance by "jumping the gun" in deploying nonspinning and replacement reserves, which resulted in suppression of market prices.

The chosen solution was to increase procurement of responsive reserves and impose stricter compliance standards, such as more frequent and random testing, on their providers. This would give ERCOT more headroom from a reliability standpoint, allowing its operators to move higher in the real-time energy bid stack than they had in 2007. The market would have a chance to clear near the top of the offer stack to set scarcity rents before ERCOT needed to take out-of-market actions.

The increase in the procurement of responsive reserves began on January 1, 2008. The ERCOT IMM reported to the ERCOT Technical Advisory Committee in early February that the chosen solution appeared to have sharply reduced the out-of-market actions by ERCOT operations and allowed the real-time energy market to function better.[45] This increase in the levels of responsive reserves could be considered an implicit mandatory insurance requirement (which is not much different from a capacity payment embedded in a bilateral contracting requirement), as loads would bear the uplifted cost for the additional resources. A parallel effort by ERCOT to improve reliability by enlisting demand-side participation through an emergency interruptible load service (EILS) was also implemented, and, as of this writing, discussion of further improvements along these lines was continuing.[46]

These actions should be considered temporary fixes that could be implemented quickly, given the prevailing software limitations and limited price-responsive demand available in the real-time market. Eventually, in a real competitive market, load response should set scarcity prices, and the PUCT has taken steps to realize that concept in the ERCOT market within the next few years.[47]

Will all of this help Texas reach the promised land of competitive electricity deregulation? ERCOT may have tight reserve margins over the next few years, so the energy-only approach will be put through its paces right out of the gate.[48] As of this writing, more than two years had passed since the decision by the commission to allow scarcity pricing in ERCOT spot markets, and on March 3, 2008, only two days after the cap was raised to $2,250 per MWh, balancing energy prices reached that level for three consecutive fifteen-minute intervals. In addition, many price spikes took place within the ERCOT balancing energy market during March–June 2008, resulting in five retail electric providers (REPs) failing to meet their financial obligations. These prices, which were far higher than allowed in any other North American markets, prompted a large public backlash and at least one legislative hearing.[49] While the trends look promising, however, the ERCOT market has yet to provide the levels of reliability by itself (that is, without the intervention of the ERCOT operator) that are seen in the Australian market. There is no doubt that the energy-only approach to resource adequacy in Texas is much closer to the economic gold standard of a commodity market than the capacity mechanisms, such as FCM and RPM, that are currently being used in the Eastern Interconnection. This mechanism should be considered only provisionally successful, however, in comparison to its Australian counterpart.

Conclusion: Why an Energy-Only Approach in Texas?

The framework adopted in August 2006 by the PUCT for market power and resource adequacy is unique in the United States. It establishes an energy-only resource-adequacy mechanism in the ERCOT market that raises the offer cap above the $1,000 per MWh prevailing in other North American electricity markets.[50] The rule increases the role of market forces in determining wholesale electricity prices and enhances the information available to market participants by dramatically increasing market transparency through prompt information disclosure.

So why was Texas the first U.S. market to develop an energy-only resource-adequacy mechanism? The authors believe that a number of circumstances that contributed to its development were, with respect to the

United States, unique to Texas. Because the PUCT was the regulator over the wholesale market, retail market, and transmission grid in ERCOT, commission staff had the freedom to be creative in developing a combination of best practices with innovative adaptations that the PUCT could implement by rule. In contrast, the New England ISO's Forward Capacity Market was a complicated solution based on a multiparty compromise among state commissions, the New England ISO, and market participants through an administrative law proceeding.

The commission staff took the approach of looking for best practices, even when they were tried in non-U.S. markets and informed by contacts with market monitors in the United States, Canada, Australia, and the Pacific Basin, and with the academic and consulting community addressing resource adequacy issues worldwide. The staff also needed to deepen its understanding of the nodal market design issues that were concurrently being debated and draw insights from them that could be applied to the energy-only resource-adequacy design.

In looking around the globe, there seems to be a strong correlation among energy-only wholesale market design, generation-friendly transmission and generation-interconnection policies, and successful retail competition. Transmission policy and generation-interconnection policy set in the early days of wholesale market deregulation in Texas in the mid-1990s allowed a flood of potential generation projects to enter the ERCOT market, with enough transmission always available to deliver the power from generators to loads without heavy congestion.[51] The successful retail market created greater pressure from retailers and industrial customers to avoid implementing a "top-down" capacity resource-adequacy mechanism in ERCOT than was present in other U.S. markets.

Texas has taken a different approach to address resource adequacy by allowing higher electricity prices to improve the possibility of adequate recovery of investment while imposing a much higher level of market transparency compared to all other markets in the United States. In addition, adequate flexibility allows refinements, if desired. A few more years of operation will provide the empirical basis for further evaluation of the effectiveness of Texas's approach to addressing the complex issue of resource adequacy.

5

Texas Transmission Policy

Jess Totten

In the 1990s, the Public Utility Commission of Texas (PUCT) developed open-access rules for the Texas intrastate transmission system to improve the terms of electric service, consistent with state policy that favored competition in the production and sale of electricity.[1] At about the same time, the Federal Energy Regulatory Commission (FERC) also adopted open-access rules, but, in a number of respects, the transmission policies of the two agencies differed. Important differences were related to the PUCT's objective of fostering competition among generating companies by encouraging market entry. PUCT policies also enhanced utilities' certainty of cost recovery and reduced the lag time between the completion of new transmission facilities and the initiation of rate recovery. These policies were instrumental in stimulating investment in new generation and transmission facilities.

The Role and Regulation of Transmission

Transmission systems have historically been an important part of a utility's electrical network, with the main function of delivering power from generators to customers. Modern utility electrical networks typically consist of a number of generators and a web of alternating-current (AC) transmission facilities. An AC electrical network commonly provides multiple paths to transmit power from generators to customers, to permit different generators

The views expressed in this paper are the opinions of the author and do not necessarily represent the views of the members of the Public Utility Commission of Texas.

to be employed as customers' demand levels change, and to provide redundant facilities when generation or transmission outages occur. Over time, utility electrical networks have interconnected to adjoining networks, improving reliability for customers and enabling utilities to support each other when problems arise with generators or transmission facilities in either system.

Such interconnection also allows utilities to engage in sales that result in a lower overall cost of power to them. Thus, if the Texas Utilities Electric Company (TU) were connected to Houston Lighting and Power (HL&P), and the marginal cost of electricity generation for TU were higher than that for HL&P during some period, both companies could benefit from a sale of energy from HL&P to TU at a price higher than HL&P's marginal cost and lower than TU's marginal cost.[2]

In earlier years, limited competition among utilities and nonutility generating companies emerged, stimulated in part by the 1978 federal PURPA law, which gave a special status to cogeneration and renewable energy, and the 1992 Energy Policy Act.[3] FERC and the PUCT adopted broad open-access rules in 1996 requiring the utilities they regulated to allow other utilities and independent generators to buy transmission service on a standalone basis.[4] The purpose of this reform was to facilitate a much greater level of competition in the sale of electricity at wholesale (that is, in sales to a utility, rather than in sales to an ultimate customer) than had previously been possible. Prior to this reform, transmission was typically not available as a separate service, and customers had to buy bundled service (generation, transmission, and, in the case of retail service, distribution) at rates that were regulated by FERC or a state commission.[5]

While the purpose of the open-access rules was to stimulate competition, their adoption for the transmission system represented a higher level of regulatory intervention in transmission issues. In particular, the rules imposed on transmission-owning utilities an obligation to provide transmission service on a standalone basis, adopt an agency-prescribed code of conduct, and disaggregate utility functions that were formerly combined. The question is, was this intervention necessary? Could additional competition in electricity have been possible without the adoption of open-access transmission rules?

Although this was not seriously debated at the time, it seems clear that the cost of duplicating the existing transmission system to facilitate competition from nonutility generators would have been high, and the public was unlikely to have willingly accepted additional transmission lines. Both issues would have been major obstacles to the construction of private transmission networks. Negotiating adequate access rights to the transmission systems operated by integrated utilities was a theoretical option, but it was one that probably would not have developed, especially with the prospect these utilities faced of losing customers to competing generating companies.

Without open-access rules, a nonutility generator with a power plant in, for instance, Houston, seeking to compete to sell power to a customer in Waco, would either have to build a new transmission line from Houston to Waco or negotiate terms with the incumbent utilities to use the existing network. A decision to build the line would impose a significant cost for the new facilities and risk public opposition to them. If successful, the nonutility generator would own an asset with little or no flexibility and would probably not benefit from the line beyond the ability to sell to the Waco customer, particularly after the sales contract with the customer terminated. Very little transmission was built on this basis in Texas, and little would likely have been built had the open-access rules not been adopted.

The other option—negotiating terms for the use of utilities' transmission systems—might have succeeded in some cases, but it is hard to see why incumbent integrated utilities would willingly have agreed to sell such access rights to a competitor. Incumbents might have had a reason to grant such rights by agreement if a nonutility generator had a realistic opportunity to bypass the utility network, but this was generally not the case. The reforms in the Energy Policy Act of 1992 and FERC's 1996 order were required because other solutions to increase wholesale competition were simply not emerging.

In adopting open-access transmission rules in 1996, the PUCT called for an independent organization to supervise access to the utilities' transmission systems. The Electric Reliability Council of Texas (ERCOT), which had previously operated as a regional reliability organization, was reconstituted that same year as an independent system operator (ISO) to carry out its new open-access role in addition to its reliability function. With the enactment of retail competition legislation in 2001, ERCOT began operating competitive markets for balancing energy, regulation, and reserve services and managing transmission congestion through a market-based system.

(In addition, the balancing energy market became a spot energy market in which generators could sell available energy, and retail providers could meet energy shortfalls.) These reforms were intended to enhance competition at the wholesale level and support the opening of the retail market in January 2002. The changes to the market and ERCOT's role in it, however, increased the level of control of the transmission systems by entities other than the owners.

Texas and Federal Transmission Policies Compared

The transmission network within ERCOT is an intrastate system under the Federal Power Act and is not, for the most part, subject to regulation by FERC. Instead, the PUCT has been responsible for developing transmission policy for ERCOT. FERC and the PUCT developed open-access rules at the same time, and both emphasized the core principle of open access to utilities' transmission systems by other utilities and independent generators to foster wholesale competition. From the outset, however, their policies differed to some degree. Among the most important differences between the Texas rules and the federal model were the Texas requirements for an independent system and market operator and for utilities to invest in the transmission facilities needed to incorporate new generation facilities into the network. The PUCT's decisions on these issues were adopted with the explicit objective of fostering competition through market entry in the generation sector.

Independent Operation of the Transmission System. Texas law and PUCT rules assign responsibilities related to transmission access to an organization that is independent of companies with a stake in the market.[6] Before the wholesale open-access rule was adopted, ERCOT was a regional reliability council. Operating under state legislation and PUCT rules and oversight, ERCOT is responsible for ensuring reliability and transmission and distribution access, accounting for generation and transmission transactions, and acting as a registry for the relationship between retail customers and their service providers. In practice, this means that generation is scheduled through ERCOT, and ERCOT manages transmission congestion; operates markets for balancing energy, regulation, and reserves; charges and pays for these services and for the relief of congestion; plans enhancements to the regional transmission system; and serves as a clearinghouse for information relating to retail customers.

Interconnecting New Generation. In many areas of the country, FERC's access rules permit transmission service providers to assess to a new generating company some or all of the costs of the new transmission facilities that are required for the safe interconnection of the generator or to allow the generator access to its preferred market.

Besides the upfront costs, the major difficulty with this FERC policy from the perspective of a generation developer is the benefit that the addition of new facilities to a complex transmission network can provide to parties other than the developer. Other companies can, for instance, take advantage of any additional transfer capacity that is created. A generator that must invest in transmission to get a reliable interconnection or the level of service it seeks is thus faced with the prospect of providing economic aid to its competitors.

The FERC has reformed its rule to provide credits to a transmission customer that pays for system upgrades.[7] Thus, if the transmission customer funds a new transmission facility, it is entitled to credits against its future transmission charges, so that the payment made by the generator is returned to it over time. Although this reform reduces the impact of the requirement that generators must pay for the new transmission facilities that they need to interconnect or provide the requested transmission service, they are still required to pay for the transmission investment in the first instance; they may still be financing improvements that will benefit other firms; and they are subject to the utility's determination of the need for the facilities and the prospect that competitive interests will affect the utility's decision concerning that need.

In contrast to the FERC policy, shortly after the initial open-access rules for the ERCOT region were adopted, the PUCT modified them to make clear that the utility to which a generator is newly connected in a transmission system is responsible for the bulk of the costs of interconnection. The Texas rule requires the generator to install a substation with protective devices and transformers, but transmission lines to connect to the electrical network and provide access to markets are the responsibility of the utility.[8]

(The generator is also required to post a deposit for any transmission improvements the utility will construct, but the deposit is returned when the generator is completed and begins producing.) This policy is based on the view that requiring the generating companies to pay for the additional transmission facilities is a deterrent to investment in new generation facilities. It encourages investment in new generation by providing for the necessary transmission by means of utility investment instead.

One reason FERC requires generating companies to be charged for these costs is that doing so forces generation developers to take into account all of the costs associated with their facilities. If a company is considering several sites for new generation, it should choose the one where total generation and transmission costs are lowest. The PUCT policy, on the other hand, means that if the generation company can ignore the transmission decisions in deciding on the site for a facility, it will invest at the site with the lowest generation-related costs, while imposing the transmission costs on the utility that installs the transmission facilities. Because the developer is not required to consider the transmission costs in making decisions about generation investment, the outcome may not be optimal from a societal cost perspective. It does, however, encourage market entry for competitive reasons. Moreover, the policy is not carte blanche for the construction of new transmission facilities. Substantial transmission facilities that are required to interconnect or provide transmission service for a generator are subject to review in the regional planning process and to PUCT approval.

The respective results of the two policies differ, as well. In many areas of the country, including the ERCOT area, ownership in the generation sector was highly concentrated when the initial open-access rules were implemented. By deterring investment in generation, the FERC policy does little to reduce this concentration, while the greater incentives for investment in new generation provided by the PUCT policy are more effective in this respect.

Furthermore, in many places the transmission system is not operated by an independent transmission organization like ERCOT but by an integrated utility instead. Such utilities are required to functionally unbundle their transmission and wholesale sales functions and operate under a FERC code of conduct,[9] but nonutility generators have generally been skeptical of the independence of the entity providing transmission service from the utility's generation operations.[10] The generation arm of an integrated utility is typically in competition with nonutility generators for wholesale power sales, and independent generators have complained that the integrated utilities act in the interest of their generation operations and against the interests of their competitors in assessing what facilities are needed to provide a competitor's requested transmission service. By establishing an independent organization to assess the capability of the transmission system and plan new facilities and by requiring that utilities bear the cost of new transmission facilities, the PUCT has removed a major point of contention between the transmission customer/generator and the transmission provider.

Clearly, the PUCT policy of encouraging new electricity generation has been a success. From 1995 to 2006, independent generators built nearly 30,000 megawatts of new generation facilities in ERCOT—roughly a 40 percent increase in capacity, most of it from combined-cycle combustion turbines. Figure 5-1 shows capacity additions, by year, in the ERCOT region.[11]

In 1995, significant levels of capacity came from inefficient gas-fired steam turbines. These units operated at higher heat rates (incurring higher fuel costs) than the new combined-cycle turbines. Their prevalence provided a market opportunity for the developers of the more efficient combined-cycle facilities, and the Texas transmission rules provided access to customers.

Initially, the new, efficient generators were able to sell power to smaller utilities, such as electric cooperatives, that were seeking to reduce their cost of buying power. With the introduction of retail competition, a larger market opened up. The new generators found new retail providers that needed power, and competition induced the incumbent providers to buy from them as well. Retail competition also resulted in entry in the generation sector by new companies through purchase of generating capacity from incumbents. As a result, roughly 10,000 megawatts of inefficient capacity was retired or mothballed between the opening of the retail market in 2002 and 2007.

In sum, several factors in ERCOT resulted in a reduction in market concentration in the generating sector: market entry by companies that had little or no presence in the region; retirement or mothballing of older generation units; and sales of generating capacity.

Unified Transmission System. In 1996, the PUCT achieved several important goals, similar to ones that FERC would seek to achieve, with limited

FIGURE 5-1
ERCOT GENERATION ADDITIONS

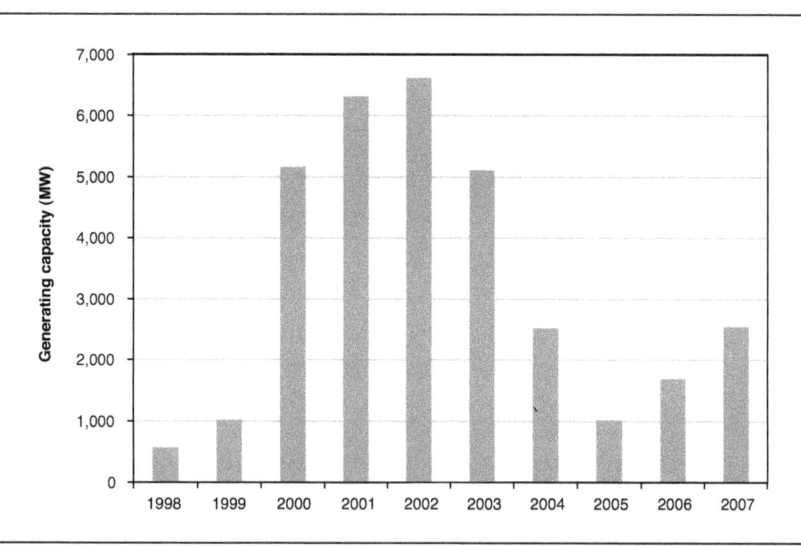

SOURCE: Public Utility Commission of Texas, "New Electric Generating Plants in Texas" (September 2, 2008).

success, with its standard market design initiative in 2002.[12] It did so by establishing uniform regional rules and directing the creation of an independent entity to manage the utilities' transmission systems so that they operated as a single system. This approach had a number of implications:

- Prior to the adoption of the wholesale open-access rules, over thirty investor-owned utilities, municipal utilities, and electric cooperatives owned transmission facilities in ERCOT, and ten utilities operated control centers to balance generation and load. The PUCT required all transmission owners to provide open-access transmission service and all transmission-owning utilities that owned generation facilities to provide ancillary services. In addition, the independent system operator was directed to create an online information system for the transmission facilities in the region.[13]

- A regional transmission rate was established that included all of the capital and operations and maintenance costs in the regional rate.[14]

- Where jurisdictional impediments kept FERC from applying open-access rules directly to municipal utilities, federal utilities, and most electric cooperatives, the PUCT's jurisdiction over the wholesale transmission service of municipal utilities and electric cooperatives[15] allowed it to impose a uniform set of rules on all transmission owners. Several large municipal utilities and generation and transmission cooperatives own transmission facilities in ERCOT, and they participate, along with other economic entities, in the energy markets operated by ERCOT and in the planning and construction of new facilities.

- FERC implemented open access by requiring each regulated utility to adopt an open-access transmission tariff. While these tariffs were highly uniform, each utility could act independently in assessing whether capacity was available to provide transmission service to other firms when requested.[16]

- Since a utility was not typically required to assess the impact of a service request on all of the other utilities that might be affected by it, its grant of service could create power-flow problems on a neighboring utility's system, for which the neighbor was not compensated. Under the FERC rules, a system of transmission-curtailment procedures had to be established to deal with the unanticipated consequences of granting service on a utility-by-utility basis. ERCOT, on the other hand, adopted market-based congestion management in 2001. (Because ERCOT is interconnected to the eastern United States through converters that allow power flow into or out of its region to be controlled, there are no "loop flows" from ERCOT into the rest of the United States, or vice versa.)

In short, the PUCT established in ERCOT an independent organization to manage the transmission network, regional rates, and open-access obligations that applied to all transmission owners, and a market-based congestion-management system. These features created the framework for vibrant wholesale competition.

Transmission Investment and Cost Recovery

Transmission planning is complex, particularly in a competitive environment. ERCOT and the utilities in its region are responsible for transmission planning, with ERCOT focusing on regional bulk system needs and coordinating the planning work of the utilities, and the utilities focusing on meeting demand in their service areas. Plans for new transmission facilities are developed to meet several categories of needs: to connect new generating facilities to the network; to meet growth in demand, particularly in metropolitan areas where economic growth has been strong; to resolve reliability issues; and to reduce congestion costs. Important elements of the ERCOT planning process include regular meetings of persons who are interested in transmission issues in the ERCOT subregions and an annual report on transmission needs that ERCOT submits to the PUCT.[17] Major new transmission facilities require the issuance of a certificate of convenience and necessity by the PUCT, and approval depends on whether there is a need for the new facilities, their expected impact on customers, community interests and existing land uses, and environmental issues.[18]

In sum, transmission investment decisions are based upon generation developers' decisions to build new generation, the ERCOT planning studies and stakeholder input, utilities' determinations with respect to needs for local service reliability, and the licensing decisions of the PUCT.

In its assessment of transmission facilities that are proposed for economic reasons, ERCOT weighs the cost of a project against the lower energy prices that will result from reducing congestion. Generation companies and retail providers have an opportunity to participate in ERCOT's regional planning meetings, so they can also present their views on the economic issues related to transmission planning.[19] Under ERCOT's planning criteria, projects that are considered on economic grounds are approved only if they provide clear benefits that exceed the costs. Over time, new transmission facilities have resulted in lower congestion costs in the region.[20]

The transmission rates in ERCOT are regional rates in which all of the costs of the transmission providers in the region are blended and charged to customers based on the peak load during four summer months; distance is not a factor. In the competitive environment, transmission providers charge the distribution companies for transmission service, and the distribution companies assess transmission and distribution charges to retail electric providers. Retail rates are not regulated, but a retail provider's cost of providing service includes the regulated transmission and distribution charges, which the provider would be expected to include in customers' prices. In the competitive areas as of 2006, the distribution companies charged retail providers about $0.006/kWh for transmission and $0.016 for distribution for a residential customer (excluding metering and other fixed monthly charges).[21]

All generators, retail providers, municipal utilities, and cooperatives have the ability to use the transmission system. Their transmission rights are determined by the rules of the congestion management system. A generator or retail provider is not assessed facilities charges when it schedules power and uses the transmission system to serve a customer for a particular day, week, or month. Retail providers' payments are based on the utilities' rates for transmission service and their customers' consumption each month. When congestion arises on the transmission system, charges are incurred by the generator or retail provider that schedules power over the congestion point, because ERCOT incurs costs in managing the congestion. The revenue from these charges does not go to the transmission providers, however; rather, it represents an additional cost of generation that is used to resolve the congestion and is paid to generators.

The transmission rules in ERCOT also include an annual adjustment in the rates for changes in peak demand, and a mechanism for annual adjustments to transmission providers' cost of service to reflect capital additions. A transmission provider that has installed new facilities can apply to the PUCT for a rate adjustment that reflects the additional invested capital and the associated depreciation, federal income tax and other associated taxes, and the return at the commission-allowed rate of return.[22] These proceedings are normally expedited and a decision rendered more quickly than in a rate case in which all of the utility's invested capital and expenses are examined, taking two or three months rather than nine or ten.[23] The PUCT's rules also allow distribution utilities to adjust rates twice a year to reflect changes in the transmission charges that they are assessed by transmission utilities.[24]

The Texas rules were intended to foster investment in new transmission facilities, both to meet the needs of a growing economy and to facilitate competition, and they have been successful. ERCOT utilities invested roughly $3.5 billion in transmission during the period 1999–2006, with 5,200 circuit miles of transmission put into service—a 15 percent increase in transmission mileage.[25]

ERCOT Development and Governance

Before wholesale open access became the rule in the state, ERCOT existed as a regional reliability council with little or no official recognition in state law and regulation. With the adoption of wholesale transmission access rules, the PUCT saw the need for an organization to oversee the operation of the transmission system and ensure that the promise of open access became a reality. Consequently, ERCOT became an independent system operator.

Over time, the organization has changed in several respects. Today ERCOT has broader responsibilities and is directly subject to state laws and regulations. The ERCOT organization is a not-for-profit corporation, so it is not subject to discipline from owners demanding a reasonable return on their investments. Texas relies on membership criteria for the governing body, "sunshine" rules, and direct PUCT oversight to ensure that ERCOT is responsive to the interests of the customers and market participants who must interact with it.

ERCOT Functions. The Texas retail competition legislation established the concept of an independent organization that is responsible for key market functions and required to have a governance structure intended to ensure its independence from any business sector in the electricity market. The functions of an independent organization under this statute are

- maintaining the reliability of the electrical network;
- ensuring open access to the transmission system;
- supervising wholesale settlement; and
- serving as a registration agent for retail providers and electric customers.[26]

For ERCOT to take on these functions, the development and approval of detailed market rules was required. ERCOT and market participants had to acquire computer and communications systems to bid and schedule energy under the new rules and to enable ERCOT to assess the status of the electrical network, and staff had to be hired to operate the new systems. ERCOT has since become the operator of wholesale markets for balancing energy, regulation, and reserves. Through these markets, the organization competitively acquires resources to balance supply and demand, manage transmission congestion, and ensure the reliability of the electrical network. The changes have resulted in a significant expansion of ERCOT's responsibilities and size, in terms both of number of employees and budget.

In addition, ERCOT manages a stakeholder process for the development and modification of policies and market rules, primarily relying on public meetings of interested persons led by stakeholder-elected committee chairmen. ERCOT also bears significant financial risks in buying capacity and energy for the regional market, the costs of which it charges back to companies that serve customers in the region. Finally, ERCOT funds and provides access to market information for an independent market monitor that is responsible for detecting violations of market rules and identifying market design improvements.

Governance. The Texas statute dealing with independent organizations was amended in 2005 to provide for more robust oversight of ERCOT by the PUCT and to require ERCOT to fund an independent market monitor reporting to the PUCT.[27] Over time, the membership of ERCOT's board of directors has changed, with the addition of retail customer representatives and independent board members. Under current law, the voting members of the governing body include five independent members, three customer representatives, six stakeholder representatives, and the chief executive officer.[28]

The PUCT rules governing ERCOT require that the meetings of the governing body and stakeholder groups be open to the public, with advance notice of the time, place, and agenda, and they require ERCOT to provide documents upon request.[29] These rules ensure a high degree of visibility of ERCOT activities and opportunities for public participation in its deliberations on market policy. ERCOT also provides continuous reporting of market outcomes and conditions. It produces reports on its operations at the monthly meetings of its board of directors; periodic reports on generation facilities under development and generation capacity, demand, and reserves; and annual reports, including a financial report and a report on transmission needs and recommendations. Separately, the independent market monitor provides an annual report on the state of the ERCOT wholesale market.

The ERCOT and Regional Transmission Organization (RTO) Experience. The ERCOT organization has evolved largely in response to directives of the Texas legislature and the PUCT, based on the needs for, first, a wholesale competitive market, and, second, a retail market. Among the factors that have resulted in change are instances of poor performance. In 2001 and early 2002, for example, ERCOT needed months to get the newly developed retail customer registration systems to operate as intended. In response, the PUCT proposed rules in September 2002 (ultimately adopted) that established requirements for ERCOT relating to public meetings, access to records, and reporting to the PUCT.[30]

Willful misconduct has also led to new directives. In 2004 it was revealed that certain high-level ERCOT officials had set up shell companies that they used to bill the organization for services that either violated its conflict-of-interest rules or represented work that was never done. These abuses involved managers who were able to award contracts for various services to companies that they owned. Following investigations by state law enforcement officials, several former ERCOT officials were convicted of various offenses and ordered to pay restitution.[31] The revelation of this misconduct prompted a number of audits to identify flaws in ERCOT's financial controls, resulting in major changes to contracting, accounting, procurement, and human resources procedures to improve the organization's security. In the 2005 session, the Texas legislature adopted amendments to the law relating to independent organizations that increased the number of independent directors and strengthened the PUCT's oversight authority. In both the instances of poor performance and misconduct, then, discovery of the problem was followed by mandates for reform from the PUCT or legislature and internal efforts to resolve the causes.

In a recent article, Michael Dworkin and Rachel Goldwasser provided a critique of the governance of regional transmission organizations like

ERCOT, raising questions about whether and to whom they are accountable.[32] The authors observed that RTOs are performing functions that are vitally important to the public, but they questioned whether the accountability to which they are held for their performance is sufficiently strict. They pointed out two problems:

- RTOs are subject to competing influences (satisfying the public interest, responding to the regulatory authority of FERC, satisfying the legitimate ends of industry stakeholders, and, in the case of multistate RTOs, responding to the concerns of the states in the region).
- RTOs are organized as nonprofit corporations, so they act without their own financial interests at risk.[33]

Dworkin and Goldwasser proposed several means of improving the governance of RTOs, including enhanced FERC oversight, stronger public interest representation in stakeholder forums, greater transparency of activities, report cards on performance, and performance-based pay for managers and directors.

Like the FERC-approved RTOs, ERCOT performs vital public functions and is subject to most of the same governance concerns as those raised by Dworkin and Goldwasser. The PUCT has adopted with respect to ERCOT some of the measures these authors suggested—namely, a high level of regulatory oversight and sunshine rules, which have resulted in greater transparency of the organization's activities. The PUCT rules also require that customers be permitted to participate in stakeholder forums. This has been only partly successful; while customer representatives have the right to participate, they may not have the knowledge and resources to do so effectively. Consequently, the representatives generally focus their efforts on the board of directors and the most important committees. A formal report card on ERCOT performance has not been adopted, and a performance pay program that was adopted was eliminated during the 2006 review of the fee the PUCT authorizes to cover ERCOT's operating costs.[34]

On several occasions, problems relating to regional organizations have been significant enough to receive public notice; these events may provide insight into factors that present challenges to the effectiveness of RTOs and

ERCOT. Probably the most prominent have been significant problems in the California wholesale market in 2000 and 2001, the poor performance of the ERCOT retail systems in 2001 and 2002, the Northeast blackout in 2003,[35] the business failure of a retail provider (Texas Commercial Energy) in ERCOT in 2003, the ERCOT fraud issues in 2004, and the 2008 dispute concerning the independence of the market monitor for the PJM Interconnection (the regional transmission organization operating in Delaware, Maryland, New Jersey, Pennsylvania, Virginia, West Virginia, and parts of several other states in the region). An ongoing concern for the ERCOT region, and perhaps for other RTOs, is the possibility that well-capitalized companies can dominate the policy discussions on changes in market design.

Some of these problems are probably grounded in the difficulty of developing operating rules and systems for competitive wholesale electricity markets. As organized markets that replace utilities' decision-making based on their own economic interests, RTOs represent a new way of operating, and they are new or radically restructured organizations. One would expect that, over time, the problems associated with novel issues and new organizations will fade, as the organizations mature and best practices are identified. Best practices also need to be carefully considered with respect to governance matters, as Dworkin and Goldwasser suggested. Are hybrid governing boards that include both stakeholders and independent members superior to boards that consist exclusively of independent members? Is greater transparency effective in engendering accountability to the public interest? Should forms of corporate organization other than nonprofits be adopted?

Some sources of governance problems may be difficult to resolve, such as policy differences among the states in a given region and between the states and FERC. A source of policy conflict that is likely to persist is the tension between reliability and markets, particularly in light of the provisions of the 2005 Energy Policy Act that gave FERC nationwide authority over reliability of the electrical network. Advanced energy markets like PJM and ERCOT procure resources needed for the reliability of the network through market mechanisms, and the conflicts that arise between reliability and market principles will now be played out both within the regional organization and between the organization and FERC.

At least one analyst has questioned the notion of introducing the profit motive as a means to improve RTO performance. Craig Pirrong suggested that it might be more appropriate for an entity that provides centralized coordination of a power market to be constituted as a nonprofit organization.[36] Pirrong submitted that a for-profit organization performing the coordination function has a motive to "chisel" with respect to reliability—that is, to cut corners on reliability to reduce costs and improve profitability. He also suggested that in the rare instances in which the market for reliability services breaks down, a nonprofit organization would more likely be able to work outside the market rules and obtain cooperation from other parties in resolving the problem.

Regional transmission organizations operate at the border between the worlds of business and public policy. This position is evident in, for example, the adoption of credit policies for participants in energy markets. In buying energy to ensure that supply and demand for electricity in a region are in balance at all times, an RTO acts on behalf of the market participants that serve customers to maintain the reliability of the customers' service. In ERCOT, if customers' demand exceeds the supply arranged by the retail provider, the provider must pay ERCOT for the amount of energy it is short at the price set in the market. ERCOT has established credit standards and collateral requirements to ensure that the retail providers have the ability to pay when they are short. Based on business considerations, an RTO might impose stringent credit standards to guard against even the most remote possibility that a company participating in the market would fail to meet its obligations, particularly if the RTO is a nonprofit organization that has limited equity as a backstop for these obligations. Business considerations might militate in favor of operating these energy markets through a for-profit organization, so it could use its own equity to provide such a backstop to the credit risks.

Competitive considerations limit how stringent the credit requirements should be for participating in the energy market of an RTO. Extremely stringent requirements might inhibit market entry and be a barrier to the vibrant wholesale competition that the RTO is intended to foster. Whether an RTO is organized as a for-profit or nonprofit entity, competitive policy would support credit policies that allow market entry on reasonable terms, and oversight by a regulatory body is probably necessary to balance the costs and benefits of a liberal credit policy.

Conclusion

The PUCT and FERC have pursued paths that are similar, if not exactly parallel, in adopting and modifying open-access transmission policy and fostering and regulating independent organizations to operate regional transmission systems and organized energy markets. Texas policy has been guided by legislation that explicitly favors competitive markets, and the PUCT has targeted its policies toward stimulating entry into the generating business in ERCOT and construction of the new transmission facilities needed for a growing economy and competitive market. Requiring an independent system operator, assigning to utilities the responsibility for transmission investment to interconnect new generators, and allowing prompt recovery of transmission investment costs were important in stimulating investment in generation and transmission in the region. By objective measures of a market's operation, these rules have been a success. Significant entry has taken place in the generation sector, less efficient generation units have left the market, either temporarily or permanently, and the costs of congestion in the region have been reduced.

ERCOT is clearly a success story, but it is less clear whether other options might be superior. The organization appears to be an important underpinning for the confidence of generators in the ERCOT market, based on its neutrality in operating the transmission system and energy market and in planning new transmission facilities. It is difficult to imagine that the level of investment that has occurred would have materialized in the absence of such confidence, but it is also difficult to assess whether other forms of organization or means of engendering accountability in an RTO would be more economical or more effective.

6

Distributed Generation Drives Competitive Energy Services in Texas

Nat Treadway

All power generation began as distributed generation (DG), with end-use loads energized by small, dispersed electric generating units located close to the point of consumption. "Distributed generation" brings to mind *technologies*; but we reject the notion that DG is merely a set of successful technologies. Technologies are successful when they provide a valued service. In the 1980s, successful cogeneration projects in Texas demonstrated that industrial customers could finance, construct, and operate large-scale generating units and sell excess power into the market, and thus set into motion electric sector reform. Today, the value of DG arises from the ability of a service provider to match a particular set of technologies and services to a particular customer's needs. DG is one of a myriad of competitive energy services now available in the Electric Reliability Council of Texas (ERCOT) market that, with enhanced infrastructure, will drive additional reform. Through its technical characteristics and ability to reduce external costs, DG serves retail customers while providing value to the electric system and society.[1]

DG may involve the application of a standard technology to a common need, such as diesel backup generators for occupant safety in buildings, or

---

I would like to thank the editors for their support and encouragement. I am extremely grateful to Jess Totten of the Public Utility Commission of Texas for his useful suggestions and corrections on matters of policy and law, and to Tommy John, Dan Bullock, and Rich Herweck for their comments on technologies and fuels. The errors and opinions that remain are solely my responsibility.

it may require specialized services. Diesel backup generators operate when the grid fails, to provide power to elevators and emergency lights. Increasingly, however, retail customers require specialized energy services relating to power reliability and quality, or the ability to make efficient use of fuels by sequentially extracting both electricity and useful heat from one fuel source. Economic efficiency is increased when customers receive specialized services that lower their costs or increase their value of service.

Modern Origins of Distributed Generation

Although the electric industry originated over a century ago with distributed generation, the economics of large power plants, coupled with government regulations, led to the growth of a vertically integrated electric utility industry. With average costs declining and DG owners unable to sell excess electricity to neighbors, the economics of DG declined as well. Rising fuel and capital costs in the 1970s then brought pressure for reform. As part of a package of energy initiatives, the federal government passed the Public Utility Regulatory Policies Act (PURPA) in 1978.[2] PURPA required state regulatory commissions to examine utility rates, planning, cogeneration,[3] and small hydroelectric generation. So-called qualifying cogenerators were new sources of electricity not owned by electric utilities that met certain standards of efficiency. The central thrust of PURPA's section 210 was to encourage nontraditional production of electricity.[4]

Dramatic growth in cogeneration projects began in Texas in the early 1980s. Texas responded to PURPA and local pressure for reform with laws authorizing utility purchases from cogeneration projects, and several policies favorable to cogeneration were adopted during the biennial sessions of the Texas legislature, in odd-numbered years from 1981 through 1987.[5]

Growth in DG energy output in Texas has largely been the result of investments in combined heat and power (CHP) technologies. To those familiar with electric power generation in Texas, CHP is also known as "cogeneration." CHP technologies generate electrical and thermal energy in a single, integrated system close to the point of customer usage. A typical CHP system consists of a prime mover to generate electricity, a heat-recovery system to capture the heat, a control system, an exhaust system, and an

FIGURE 6-1
COGENERATION CAPACITY ADDITIONS IN TEXAS, CUMULATIVE MEGAWATTS, 1920–2009

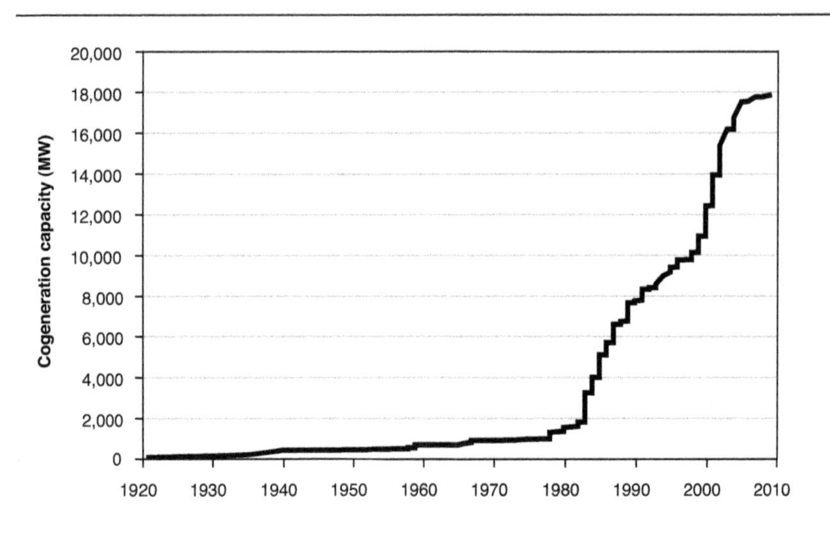

SOURCE: Prepared by the author with data from the Gulf Coast CHP Application Center (www.gulfcoastchp.org) and the Texas CHP Initiative (www.texaschpi.org).

acoustic enclosure. Of the CHP capacity in Texas, 94 percent serves chemicals and petroleum refining, and 88 percent of the electricity from DG is generated with natural gas fuel. Figure 6-1 displays the cumulative capacity additions of CHP in Texas during the twentieth century. CHP now provides more than 20 percent of the total electric generation in the state, as shown in figure 6-2. The reorganization of electric utilities and the changes in generation asset ownership account for the dramatic changes in the sources of electric generation from 1999 to 2003.

Distributed Generation Technologies and Potential

Distributed generation in Texas has relied upon numerous standard technologies that are combined into systems to address customers' need for electricity and thermal energy.

FIGURE 6-2
SUPPLY OF ELECTRICITY GENERATION IN TEXAS, 1990–2007

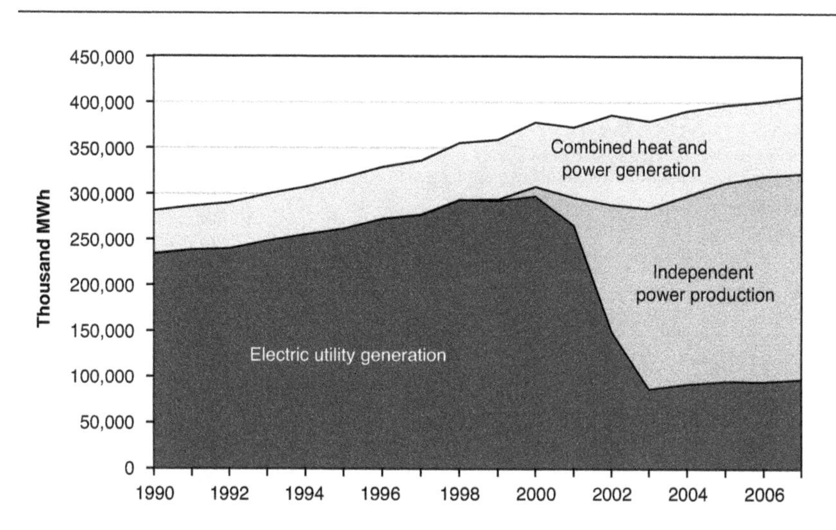

SOURCE: Prepared by the author with data from the USDOE Energy Information Administration (http://www.eia.doe.gov/cneaf/electricity/st_profiles/sept10tx.xls).

Definition of Distributed Generation. Distributed generation consists of electricity generation located close to end uses and energy customers. DG ranges in size from greater than a thousand megawatts, down to tiny generators of ten kilowatts that provide occasional emergency power to homebound individuals with medical needs, and even to solar-electric applications of one kilowatt.[6] DG is dispersed throughout the electric system, is typically located on commercial and industrial customer premises, and is designed primarily to serve the local customer load. Depending on the design of the DG system (the prime-mover technology, its fuels, its controls, and its integration into customer operations and with the grid), DG units may operate very infrequently (to provide power during emergencies) or continuously (to provide economical power, enhanced reliability, power quality, or other attributes and energy services).

We focus in this chapter on high-efficiency CHP because of its significance in providing electrical energy to Texas. CHP technologies are characterized by significant investments (high capital costs) that are economical over the long run because fuel inputs are used in a highly efficient manner

FIGURE 6-3
SCHEMATIC REPRESENTATION OF CONVENTIONAL GENERATION AND CHP ENERGY FLOWS

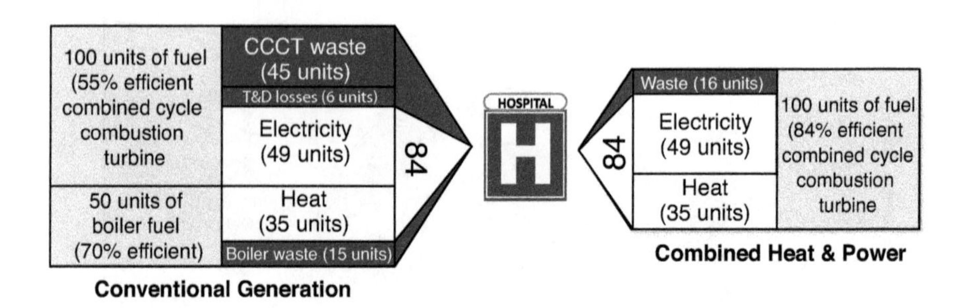

SOURCE: Author's illustration.

(low operating costs). Electricity production in physical proximity to both electrical and thermal end uses (such as refinery operation or hospitals) provides opportunities for technical efficiency. The recovery and use of heat is feasible and economical when distribution pipelines are short. Use of the heat that would have been wasted provides high thermodynamic efficiency and low per-energy-unit operating cost.

To illustrate the difference between conventional generation and CHP, figure 6-3 presents two ways to meet the electricity and heat needs of a hospital. Conventional generation (energy flows shown on the left) requires 100 units of fuel energy to generate 49 units of electricity in a distant power plant (assumed here to be a highly efficient combined-cycle combustion turbine). The hospital requires an additional 50 units of fuel energy to provide 35 units of useful heat. A total of 150 units of fuel provides 84 units of useful energy to the hospital, with the remaining 66 units wasted. The flows on the right of the diagram present the CHP approach. A total of 100 units of fuel energy are used in an integrated, sequential process to generate 49 units of electricity onsite and then capture 35 units of heat to address the hospital's heat needs. The loss is 16 units. In both cases, the building uses 84 units of useful energy. The conventional fuel input is 150 units, however, while the CHP approach requires only 100 units.

Distributed Generation Technologies and Applications. DG has a long history in Texas. More than two dozen CHP units were constructed in the state between 1920 and 1980 to take advantage of industrial customers' needs for the simultaneous use of electricity and steam. Gas turbine CHP systems were installed at major refining and chemical facilities from the 1950s to 1970s. CHP is economical in petroleum refining, production of industrial organic and inorganic chemicals, and the paper and food industries. During World War II, several unique arrangements among industry, the federal government, and electric utilities served Texas's urgent need to increase the output of both war materials and electricity.[7]

The dominant applications of DG in Texas have distinct values for electric customers and markets. First, there are direct benefits to end-use customers who install and operate DG. Second, indirect benefits accrue to electric system customers. DG has become part of the complex array of energy services available to retail customers, and customer needs drive the mix of DG technologies and services that dominate the market.

In Texas, four modes of DG operation define the marketplace: Backup generators operate infrequently and allow safe building operation during electric system outages; peak clipping generators operate occasionally for economic reasons, such as to lower energy costs or demand charges during high-use and high-cost periods; primary power generators operate most hours of the year to provide premium levels of reliability and power quality (pure waveform); and CHP units operate most hours of the year and make efficient use of fuels to provide electricity and heat (such as industrial-process steam and hot water).

A particular set of DG technologies are selected to satisfy a customer's particular needs. The simplest tradeoff relates to costs: low-first-cost units with high operating costs are economical as backup generators; high-first-cost units with low operating costs are ideal for generators that operate most of the hours in a year. The common technologies include diesel engines, natural gas engines, natural gas turbines, and steam turbines. Less common are fuel cells, which are commercial in limited applications but have great potential to lower emissions.

The technologies in table 6-1 are presented in simple terms. The particular requirements of an industrial or commercial facility would require customized engineering that would consider much more than the prime-mover

TABLE 6-1
COMPARISON OF DG/CHP TECHNOLOGIES

	Diesel engine	**Natural gas engine**
Electric efficiency (LHV)	30–50%	25–45%
Part load	Best	OK
Size range (MW)	0.05–5	0.05–5
CHP installed cost ($/kW)	800–1,500	800–1,500
Start-up time	10 seconds	10 seconds
Fuel pressure (psi)	< 5	1–45
Fuels	Diesel, residual oil	Natural gas, biogas, propane
Uses for heat recovery	Hot water, LP steam, district heating	Hot water, LP steam, district heating
CHP output (BTU/kWh)	3,400	1,000–5,000
Useable temperature for CHP (°F)	180–900	300–500

SOURCE: Adapted from *Distributed Generation: The Power Paradigm for the New Millennium*, edited by Ann-Marie Borbely and Jan F. Kreider (Boca Raton, FL: CRC Press, 2001), table 10.1, 274. Originally from Onsite Sycom Energy Corporation, draft report to the California Energy Commission: Market Assessment of CHP in the State of California (1999).

technologies. A vast array of other technologies is required to integrate a DG or CHP unit into a customer's operations, as well as to control the technology, manage its emissions, and interact effectively with the grid.

There are other important DG applications and technologies that do not fit easily into the four basic modes of operation. For example, utilities use small peaking generating units to address weaknesses in the distribution system that result from rapid growth in demand or failures from accidents caused by weather events. These generating units are generally moved from one location on the distribution system to another after the utility has addressed the weakness or made the repairs through a distribution line or substation upgrade.

Waste fuel technologies for dispersed power generation make use of waste materials that ordinarily require costly disposal. In appropriate circumstances,

Gas turbine	Microturbine	Fuel cells
25–40% (simple) 40–60% (combined)	20–30%	40–70%
Poor	Poor	—
3–200	0.025–0.25	0.2–2
700–900	500–1,300	> 3,000
10 min. – 1 hr.	60 seconds	3–48 hours
120–500	40–100	0.5–45
Natural gas, biogas, propane, distillate oil	Natural gas, biogas, propane, distillate oil	H_2, natural gas, propane
Heat, hot water, LP-HP steam, district heating	Heat, hot water, LP steam	Hot water, LP-HP steam
3,400–12,000	4,000–15,000	500–3,700
500–1,100	400–650	140–700

disposal costs can be eliminated, and the waste is transformed into a useful fuel. For example, forest industry wastes, agricultural wastes, and landfill gas can be burned to produce electricity. Forest and agriculture wastes are not generally located close to large electric loads, but the cost to transport electricity is generally lower than the cost to transport bulky wastes. Small, dispersed, waste-fuel generating units will grow in importance as fossil-fuel prices rise and improved design allows better use the waste stream.

Small-scale and dispersed renewable-resource applications include small photovoltaic panels (PV or solar cells), small wind turbines, and small hydroelectric generators. Individual, widely dispersed wind turbines are considered DG, but concentrations of wind turbines are not.[8] Small, renewable DG units may receive net metering treatment under federal and state law and are the subject of targeted projects.[9]

The future holds potential for the integration of vehicular power within the stationary electric system. It is straightforward to project that electric vehicles will become substantial new loads to the electric industry because of the need to recharge their batteries. The batteries may eventually provide substantial energy storage, peaking power, and ancillary service as well.[10]

Opportunities for the Application of DG Technologies. The greatest opportunities for DG in Texas fall within the four major applications: backup generators, peak clipping generators, primary power generators, and combined heat and power generators.

Backup generators are installed whenever local building and safety codes require them, and they operate for a few minutes or hours during the year. In many instances, operation is limited to the required periodic safety testing of a diesel generator. The impact on the environment of testing and limited operation is small, and the output is negligible. The potential for growth in backup generation is in direct proportion to economic and population growth. The most interesting development is related to the control of backup units as a source of electric capacity or demand-responsiveness. Several companies specialize in this market.[11]

Peak clipping units operate from a few hours to several hundred hours per year to avoid high-use and high-cost periods. Peak clipping generators focus on cost control and could be combined with demand-response programs. Because they are designed to operate from dozens to hundreds of hours per year, they are subject to more stringent environmental regulation than backup generators.[12]

Primary generators and high-efficiency combined heat and power units both provide power in a manner designed to satisfy the customer's needs for cost control, improvement of power quality, and management of volatile fuel costs. Primary power generators are associated with data centers and certain manufacturing processes that require high levels of reliability and power quality. Because of the cost of operation and the high value of manufacturing, these DG applications may not export power, but focus on critical power needs.

The source of new DG electricity production in Texas will overwhelmingly be combined heat and power technologies. CHP units are designed to use fuel with great efficiency and operate all year round. Generally operating about 90 percent of the time, or nearly 8,000 hours per year, these individual multipurpose generating units recover or reuse heat. Customers may own multiple units to have year-round reliability and build redundant capacity. Most CHP installations in Texas are large. Only forty-five installations of less than 10 MW have been built in Texas, representing less than 2 percent of existing CHP capacity.

While market penetration of CHP is already significant, industrial potential for its use remains in refining, chemical, food, and paper industry applications. Industrial-sector potential in Texas is estimated at 7,411 MW, and commercial- and institutional-sector potential at 6,229 MW.[13] Of the commercial and institutional potential, 76 percent will be found in hotels, hospitals, colleges, schools, office buildings, prisons, and nursing homes, with the remainder spread among other segments. The Gulf Coast Regional CHP Application Center has determined that 75 percent of the CHP potential can be identified in applications of less than 20 megawatts each, divided evenly among applications of less than 1 megawatt, of 1–5 MW, and of 5–20 megawatts.

There is a great deal of CHP to be developed, but it will take a concerted effort to work with a large number of small to midsized customers. Significant challenges will also lie in identifying and capturing for economical use the existing major sources of waste heat and opportunity fuels.

Institutions That Affect DG and Competition

In Texas, the ready availability of suitable cogeneration sites in the 1980s gave impetus to customer investment in cogeneration and subsequent market reforms. Texas regulators, focusing on how to acquire low-cost, reliable electricity, compared power from cogeneration on a more or less equal footing to that from central generation power plants. In doing so, they gave cogeneration an opportunity to enter the market in a substantive manner, with their administrative process of comparing alternatives setting a framework for market-based comparisons. In the 1980s, electric utilities secured long-term bilateral power contracts with qualifying cogenerators to lower the cost of electric service and meet peak capacity needs. By the 1990s, the utilities were seeking regulatory guarantees for stranded-cost recovery based on the loss of sales to qualifying cogenerators and in anticipation of additional losses in the future. The formal request for stranded-cost recovery was a clear admission that the economies of scale that had been used to justify ever larger central generation power plants for certification and cost recovery were no longer particularly relevant, as they rested on the proposition that existing assets were more costly than planned alternatives.

The reforms that enabled retail competition took place over thirty years. They began in 1975 with the creation of a state agency to regulate electric utilities, and they continued through the late 1990s with the implementation of open-access transmission and competitively secured contracts for firm power. They go on today, as regulators determine the degree of unbundling of distribution services. The regulators' decisions affect the breadth of choices available on customer premises—that is, the breadth of competitive energy services available to retail customers.

Texas Embraces an Administered Market for Cogeneration. The creation of the Public Utility Commission of Texas (PUCT) in 1975 opened the door for scrutiny of electric utility operations and rates and utility resource planning. Prior to the Public Utility Regulatory Act of 1975 (PURA 1975), electric utility ratemaking was regulated by local municipal governments through franchise agreements. Such an agreement allowed the electric utility to use local public rights-of-way for its distribution system, provided it charged reasonable rates for electric service and paid a franchise fee. As long as electricity rates were declining, service was "adequate and efficient," and rates were "fair, just, and reasonable," this system worked without great controversy.[14] Since costs had been decreasing, utility planning had been successful, although it was immature.

In the 1970s, the system began to unravel. As fuel costs rose, local governments resisted rate increases, and the system became unworkable. Utilities needed more, not less, certainty of cost recovery if they were to expand their systems to meet growing needs for power in Texas. Because they served multiple communities, certainty for them required state action. The sixty-fourth Texas legislature adopted PURA 1975 and gave the new PUCT ratemaking authority. PURA required a utility to get a special permit—a certificate of convenience and necessity (CCN)—to construct a power plant facility. Once acquired, the CCN increased the likelihood of cost recovery for the associated asset.

The federal adoption of PURPA in 1978 and Texas laws affecting cogeneration in 1981 provided the basis for CHP investment in Texas. From the early days of industry in Texas, the state's industrial base presented favorable economics for the joint production and use of electricity and steam along the Gulf Coast, notably in and around Houston–Galveston–Freeport, Beaumont, and Corpus Christi. The Texas legislature gave the PUCT additional oversight responsibilities for long-term utility planning. Starting in 1983, the PUCT mandated a two-step review process for power plants (a notice of intent, followed by the CCN), and began to require utilities to consider resource alternatives to the construction of new power plants. The PUCT then began to administer a market-cost test to long-term planning decisions as part of its CCN review. The market-cost test evolved from an administrative calculation in the 1980s to an administered market[15] in the early 1990s. Full bulk-power competition in the late 1990s reduced the need for a regulatory cost test.

Texas defined avoided capacity costs in the immediate aftermath of PURPA through a negotiated compromise that came to be known as the "three-dollar docket."[16] In addition, energy exported to the transmission grid was reimbursed at a rate equal to the utility's short-run marginal costs (fuel and variable operations and maintenance costs). In 1984, the PUCT developed the *Texas Cogeneration Manual*, which reduced uncertainty with respect to the rules of the administered market. The PUCT also adopted criteria for the calculation of avoided costs, the consideration of cogeneration in plans for expansion of utility capacity, and the need for formal certification of any cogenerator or small power producers that wanted to sell power to retail customers.[17] Also in 1984, the PUCT considered a consultant report that recommended the committed-unit-basis methodology for the calculation of avoided costs.[18] Under the committed-unit-basis approach, the avoided-capacity payments to cogenerators faced a price ceiling based on the generating unit that would have been constructed by the utility if it did not purchase cogenerator power. Power contracts between utilities and cogenerators with prices below the avoided capacity cost were considered prudent.

In 1985, the PUCT determined that cogenerators shopping for high buyback rates (that is, high avoided costs) were entitled to have their power "wheeled" (that is, transferred across the transmission wires) by the local utility to other utilities that offered more attractive buyback terms. Wheeling of power across the transmission system established a mechanism for use of the transmission grid by nonutility power producers.[19] This decision was supported by the Houston Lighting and Power Company, which anticipated being swamped by applications from cogenerators to force it to buy power under the terms of PURPA.[20]

By 1986, routine avoided-cost proceedings were established to examine a utility's plans for capacity expansion, define the generating unit that could be avoided through the use of cogenerated power, and calculate the value of avoided costs (energy and capacity payments) associated with the avoided generating unit. Utilities could build on their experiences with power purchases from one service territory to another. Now it became more common for a utility with capacity-resource needs to negotiate a long-term purchase agreement with a qualifying cogenerator in another utility's service territory. The calculated avoided energy and capacity payments formed a ceiling for payments to cogenerators.

During this same period, several of the state's investor-owned utilities were experiencing delays in nuclear power plant construction, along with financing difficulties and a continuing need for electric capacity. These utilities forecasted a demand for coal-fueled generating units and presented favorable capacity payments for cogenerators. Other utilities that were trying to limit the expansion of the rapidly growing cogeneration industry in their service territories sought low avoided capacity costs (associated with simple-cycle combustion turbines) to move negotiations in their favor.

In 1987, the Texas legislature affirmed the inviolability of power contracts between cogenerators and utilities and required the PUCT to treat requests to construct transmission lines expeditiously so that cogenerators would have greater accessibility to markets.[21] Texas added cogeneration capacity and brought the benefits of high-efficiency generation to ratepayers and industrial cogenerators. Step by step, the Texas legislature addressed the principles set forth in PURPA, and DG construction accelerated rapidly.

Texas Embraces Transmission Access and Competitive Generation. In Texas, the development of DG alternatives challenged the traditional electric-utility model of power generation and delivery. The rapid expansion of cogeneration in the 1980s provided an impetus for wholesale competition reforms in 1995 and for retail competition reforms in 1999. DG today continues to challenge bulk-power market notions about what makes a market fully competitive. Onsite power generation provides many services of value to electricity customers, some of which are described below.

Cogeneration Drives Market Reform. Cogeneration system owners—typically large, industrial customers—have been instrumental in demonstrating how restructuring is both possible and desirable. Initially, two dozen self-generators operated in Texas. As the base of independently owned and operated generating units grew, they became a political force. Self-generation provided a cost benchmark for the electric industry as a whole, and surplus electricity in search of a market brought transmission access to the forefront of negotiations and legislative reform.

Not everyone can install DG, but every retail customer in Texas benefits from lower costs when an electric utility purchases low-cost power from a cogeneration project. Electricity customers benefit from the diversification of the physical locations of generation and from a diversity of generation technologies and fuel sources. Through widespread use of DG, the ownership of generating resources is diversified and dispersed, increasing the number and types of market participants. Increasing the number of power producers should enhance market performance by reducing the market power of any one player. Electricity customers also benefit from enhanced reliability and security through the diversification of resources on the network, and from the value provided by DG customers who are able to respond to market prices and emergency signals from the system operator. DG reduces the load on the transmission and distribution system, which reduces line losses immediately and defers the need for line upgrades over the long term.

Large cogeneration projects in Texas have also demonstrated that nonutilities can finance, construct, and operate large-scale power generating units. Industrial customers (or developers) who own CHP assume significant risks. The risks associated with a large power plant include construction delays, cost overruns, and fuel price increases. Nonutility power plant developers assumed these risks at a time when several of the large electric utilities in Texas were dealing with nuclear power plant delays, cost overruns, and capital cost increases. The value of ten-year power contracts from cogeneration development was a welcome addition to a state that was concerned about nuclear plant cost increases and delays.

Federal and State Reforms. The Energy Policy Act of 1992 (EPAct 1992) enabled greater competition in the provision of generation services by allowing independent nonutility power producers to participate in power markets.[22] The success of CHP in providing low-cost firm power without jeopardizing the transmission system gave strength to arguments in favor of regulatory reform.

In 1995, Texas enacted legislation to address wholesale competition in the electric industry, including matters addressed in EPAct 1992 and by the actions of the Federal Energy Regulatory Commission (FERC). PURA 1995 focused on open-access transmission, comparability of transmission service for nonutility power producers, and requirements for utilities to conduct resource solicitations.[23]

The 1990s brought further reforms. Changes in the structure of the wholesale market increased access to bulk power markets, and construction by nonutilities increased markedly. Increased market volatility and fuel prices and a greater need for environmental protection set the stage for more onsite generation, customer cost control, and public policies to encourage efficient power production, increased use of waste fuels, and renewable energy investments.

Texas Embraces Distribution Unbundling and Competitive Energy Services. In Texas, the emergence of a competitive retail market was not merely an opportunity for resellers to aggregate customers and repackage traditional electric service options. The law anticipated that retail electricity choice would enhance innovation and result in new services.

Innovative technologies and services may change the way customers interact with the utility network and with their devices, or fundamentally alter how energy services are provided. Innovation also provides opportunities for offering a range of products and services that satisfy customer preferences. Choices relate to the physical attributes of electricity as well the convenience of the customer in using and paying for it. They include

- choices relating to the time of usage;
- choices relating to reliability and quality (such as power firmness, voltage fluctuation, the ability to curtail usage, and the ability to obtain backup service);

- choices relating to geography (point of generation to point of delivery);
- choices relating to power service (addressing fuel and technology preferences, such as green pricing); and
- choices relating to contract terms, price-risk management, service guarantees, and maintenance contracts.[24]

The laws that govern the interactions among customers, electric utilities, and competitive service providers are very important to efficient market outcomes. Texas law now separates the network that provides transmission and distribution service from electric retailing and competitive energy services. Retail electricity customers in the competitive portions of ERCOT (those not served by municipal or cooperative utilities) who are not fully satisfied with power contract alternatives or the reliability or quality of power delivered from the grid can consider other options. Customers served by municipal or cooperative utilities or located in the non-ERCOT region of Texas have similar onsite options. Onsite generating units can be designed to satisfy a customer's exacting needs with respect to power reliability, power quality, risk management, and the level of integration into the customer's industrial processes or commercial facilities. Onsite generation may also satisfy qualitative needs related to corporate image and branding.

Innovation drives markets to efficient outcomes because customers can address concerns about particular attributes or dimensions of service. In a regulated market, the state regulatory authority, municipal utility, or electric cooperative board of directors defines a package of services and sets the retail prices, attempting to do so in a consistent and open process. While this consistency and openness is essential to the deliberative political process, it does not serve an individual's unique needs. A competitive retail market allows entrepreneurs to innovate and take risks with new technologies and services that do address customers' needs. While DG is technically possible in many settings, entrepreneurship can flourish where there is retail choice because the constraints (such as administrative burdens) of a tariff book are eliminated.

As it prepared for a competitive marketplace, the PUCT took seriously the need to unbundle and separate electric utility facilities and functions. As early as 1996, with its adoption of integrated resource planning (IRP) rules, the PUCT stated that all resources should be competitively acquired by vertically integrated electric utilities. Competitive acquisition meant that the utility would issue a request for proposals, including competitive bids from the developers of independent power projects, CHP projects, and demand-side management resources. Adoption of the IRP rule expressed a general policy on the part of the PUCT for the functional unbundling of distribution service to separate monopoly functions from competitive services. The purpose was to increase opportunities for customers and for service providers, and to minimize the potential for utilities to engage in anticompetitive behavior in the energy service sector.[25] Competitive energy services—essentially all premises-based activities—were recognized as being solely within the competitive sphere and beyond the scope of regulated distribution utility services.

In 1998 and 1999, the PUCT conducted a rulemaking proceeding to define what customer services should be competitively provided and what services should continue to be provided by the utility. The commission defined "competitive energy service" as comprising "customer energy services business activities that are capable of being provided on a competitive basis in the retail market."[26] These included DG, among many others.[27] The PUCT further stated that an "electric utility shall not provide competitive energy services."[28] The new regulations reflected the understanding that electric utilities had a history of interacting with retail customers in ways that were not always conducive to the interests of competitive service providers.[29]

The PUCT required utilities to separate costs into four categories: customer service (including metering, billing, and interaction with customers), distribution service, transmission service, and generation service.[30] The cost separation allowed utilities to recalculate rates in narrower categories and to begin eliminating cross-subsidies. When the unbundling decisions of 1998 were incorporated into amendments to PURA 1999, the effect was the codification of the formal separation of the wires functions (transmission service and distribution service) from electricity retailing and competitive energy services functions. PURA 1999 required each utility to separate regulated activities from "its customer energy services business activities that are otherwise also already widely available in the competitive market" by September 2000, and to separate its business activities into "the following units:

(1) a power generation company; (2) a retail electric provider; and (3) a transmission and distribution utility" by January 2002.[31]

With the separation of regulated from unregulated electric utility functions, the nonutility competitive providers were given a stable platform from which to serve customers. The crucial determinant was that competitive service providers need not worry about electric utilities using their monopoly position to offer competitive services below cost, or to tie a competitive service to a monopoly service in a way that would create a barrier to competitive providers. Unbundling of distribution services from competitive energy services reduced regulatory uncertainty and freed retailers from concerns about utilities' behavior.

Project Development Issues

Developers of distributed generation projects face a large number of uncertainties relating to fuel price volatility, regulatory costs, equipment delivery and performance, construction management, and project finance. Many of these uncertainties are unavoidable, and they present the entrepreneur with management challenges. Some of the more specific issues are discussed below.

Project Finance and Economics. Central among the economic and financial issues facing developers is the management of fuel price uncertainty and the securing of financing for CHP projects.

Uncertainty of Natural Gas Prices. In recent times, natural gas prices have remained high and volatile relative to historical levels, and onsite generation has competed at the margin with natural gas–fueled combustion turbines. High natural gas prices present an economic barrier for DG units that use fuel inefficiently; they should, however, benefit CHP installations that use fuels extremely efficiently. This presents a challenge for the economics of CHP, particularly when prices are high, because CHP competes not merely with central gas–fueled units, but with all central generation, including solid fuel units and nuclear units.

CHP developers and customers must consider the tradeoffs among alternative price scenarios and firm power offers. CHP provides a physical hedge that may be more valuable to the developer, when all issues are considered, than a contract that contains cost-escalation clauses.

Project Finance. Specialized manufacturing facilities that require high levels of power reliability and quality accept the cost of premier power installations. Onsite power generation is an option that is familiar to them. Since energy-efficiency investments have not traditionally been considered a primary objective for typical commercial, institutional, or industrial facilities, they often require special financing arrangements.

In Texas, state government facilities allow energy-efficiency projects to be financed out of the savings associated with them. The "Loans to Save Taxes and Resources" (LoanSTAR) program uses a revolving loan mechanism, initiated in the 1980s from oil-overcharge monies, to provide demonstration projects in building energy efficiency. The loans reduce costs and risks for building managers in state agencies, school districts, higher education, local governments, and hospitals.[32]

Business customers do not have similar options. Finance is challenging for the business project developer because CHP projects are often too large to finance out of cash savings, and yet may be too small to permit efficient financial transactions. The investment community is often seeking projects of $50 million to $100 million or more to keep transaction costs to a small percentage of total cost. A common observation is that it takes equal effort to finance a $5 million project as it does a $50 million project. The transaction costs of small projects comprise a greater percentage of overall costs than those of large projects; thus, these project are ripe for a standardized approach. To address this issue, a developer may aggregate the needs of similar projects and seek funding for a larger amount.

Energy-Efficiency Incentives. The Texas legislature mandated a standard-offer program for energy efficiency. As customers are not generally accustomed to making investments with long-term rewards as a result of energy-efficiency, a PUCT rule requires electric utilities to administer the process, and it provides them with performance incentives to exceed the legislative goals. The PUCT also adopted rules that make small CHP project owners eligible to receive energy-efficiency incentives. Such incentives may play a critical role as a source of capital for CHP.[33]

Interconnection and Infrastructure. DG units that connect to the grid must comply with established interconnection procedures. Small-scale DG projects are affected by policies regarding net metering and the availability of detailed interval usage data.

Interconnection Standards. Texas was early among the states in adopting DG interconnection rules in 1999.[34] Since that time, electric utilities have been required to file annual reports at the PUCT identifying each DG facility interconnected with the utility's distribution system during the preceding calendar year.[35] FERC has developed interconnection standards for small generating units of less than 20 MW.[36] Although interconnection is a less significant barrier than it was a decade ago, legislation passed in 2007 encourages distributed renewable generation—that is, renewable generation of less than 2 MW installed on the retail customer's side of the meter—so that onsite generation of all sizes can participate in the market.[37]

Advanced Metering. "Advanced meter infrastructure" refers to the use of electric meters that are capable of collecting, storing, and communicating information about electricity usage over relatively short periods of time (for example, fifteen-minute periods for energy settlement by ERCOT). Two-way communication is important not just to allow the retrieval of usage information, but also to convey control signals for demand-response initiatives and, ultimately, to control DG units and other devices on the customer premises. Larger DG projects are required to use interval data recorders (IDRs) to settle the energy production in a manner consistent with the ERCOT settlement procedures. Advanced meter infrastructure presents no new challenges or uncertainties for larger DG because existing procedures provide sufficiently detailed information. For small renewable power production and net metering, advanced meters are important because without them, these resources must rely on typical load profiles of usage and production rather than on actual measurements. Metering was the subject of a PUCT rulemaking proceeding for small-scale DG projects.[38]

Distribution Service Rates and Regulation. The setting of rates for electric utility service to support DG has traditionally been quite contentious, and it remains so in the portions of Texas in which services are not fully unbundled. With the unbundling of services in the competitive portions of ERCOT, rate issues are not as important as they once were because the PUCT ratemaking authority is limited to the costs for transmission and distribution service. Individual customers can obtain the power services they desire at market-based rates.

Standby or Backup Rates. DG customers may rely on power from the electric grid when the DG unit fails (an unscheduled outage), when it is maintained (a scheduled outage), or when it is insufficient to provide all the customer's power requirements. In Texas, backup, maintenance, and supplemental power were provided by utilities to qualifying cogenerators during the period of the PUCT's administrative oversight.[39] With the emergence of competitive markets, DG customers make arrangements with retail electricity providers to acquire what they need. A discussion of standby and backup power in traditional markets is provided by Dismukes and is relevant to the portions of Texas served by municipalities, electric cooperatives, and investor-owned utilities outside of ERCOT.[40]

Variable versus Fixed Charges for Delivery Services. Reform of electric utility regulation presents an opportunity to change the incentives for cost recovery. Electric utilities would prefer stable revenues, and they want to reduce the risks associated with revenue volatility. One way to do so is to move as many costs as possible from variable charges—related to customer usage, for example—to fixed charges, collected as per-customer monthly fees.

A movement toward fixed charges would more firmly establish distribution as a fixed set of monopoly services for which everyone must pay. This point of view contends that the costs of the distribution facilities are the same, whether or not a customer has DG and whether or not a customer is intensively using a fixed asset. A volumetric rate could result in failure to recover the full costs of serving a customer with a DG unit. Every administrative decision regarding minimum service levels imposes costs on customers who may desire lower levels of service. A DG customer with significant onsite investment is less interested in the high level of reliability provided by the distribution utility. Fixed charges impose the same cost on everyone, which makes economic sense if everyone uses the same services.

DG customers argue, however, that since their usage of the network is different, their cost obligations should be less. Many DG customers are less interested in a network with high levels of reliability than are customers who rely on the network for all their electricity needs. The administrative choice of recovering distribution costs through fixed or volumetric prices therefore affects the economics of DG.

Menu-of-Service Pricing for Delivery Services. One means of addressing variability in the choice of regulated services stands in stark contrast to the move just described toward fixed charges. Menu-of-service pricing increases the choices available to retail customers by altering the way in which distribution services are purchased. Aside from benefiting customers, more choice in the use of the transmission and distribution systems provides feedback to electric utilities on what services are needed and what customers are willing to pay for them. In Treadway's words,

> Customers no longer have a universal desire for distant power plants or for the wires that transmit this power. Customers have a few choices today, and in selecting energy services they ought to be charged based on what they want, when they want it, how much they use it, and the quality of service that they receive. Each of these dimensions—what, when, how much, and quality—has a corollary with respect to energy services.[41]

Getting feedback from customers on price and usage could affect the planning and design of the electricity network and increase risk for electric utilities. Menu-of-service pricing would require more comprehensive unbundling of the elements of service, separate pricing of each element, and increased control and communications to ensure that the utility would limit customers to the services they request. Each DG customer would determine the optimal mix of transmission and distribution services and DG assets, and customers would be able to supply their own services to supplement those they buy from the electric distribution utility.[42] A menu-of-service pricing regime would use advanced metering, communications, and controls to adjust the level of service to the preferences of particular customers. Greater utility costs would be incurred by customers who desired premium utility services, and DG customers would determine what level of competitive service to buy to replace the utility services.

Environmental Regulation and Externalities. Environmental regulation attempts to protect the natural environment by identifying the sources of pollution and restricting human actions that cause it. It takes the external costs of pollution and internalizes them by imposing them on polluters.

Emissions Regulation for DG in Texas. Air emissions in Texas are regulated by the Texas Commission on Environmental Quality (TCEQ). The TCEQ implements the Texas State Implementation Plan (SIP)[43] to remedy nonattainment of the national air quality standards set forth in the federal Clean Air Act and its amendments.[44] The U.S. Environmental Protection Agency (EPA) also has a growing presence in Texas as it works with state agencies, local governments, and businesses to address matters of national and international importance.

Air quality is a particular concern in and around large metropolitan areas in Texas. Dallas–Fort Worth and Houston–Galveston are nonattainment areas for certain pollutants, necessitating close scrutiny and increasingly stringent regulation of air emissions. Many other large cities and industrial development regions are concentrated in the central corridor and eastern portions of the state, giving rise to a distinction between east and west Texas for treatment of air emissions regulations for new electric generating units.

Owners or operators of electric generating units that meet certain stringent air quality standards may apply for a simplified Air Quality Standard Permit for Electric Generating Units.[45] Generating unit projects that do not qualify for simplified treatment may apply for approval under a "permit by rule" process. A permit is required if the unit generates electricity for more than three hundred hours per year.[46]

To ensure consistency in the way it treats generators, Texas needs to change the way it regulates air emissions. At present, some engines and turbines receive very different treatment, depending on whether they generate electricity or use shaft power directly from an engine. The same emissions standards are applied to small power plants as to very large ones, ignoring the inherent differences in energy efficiency. Small onsite generators provide many services that are not provided by central generating units.

FIGURE 6-4
ANNUAL RENEWABLE ENERGY GENERATION IN TEXAS, MEGAWATT-HOURS

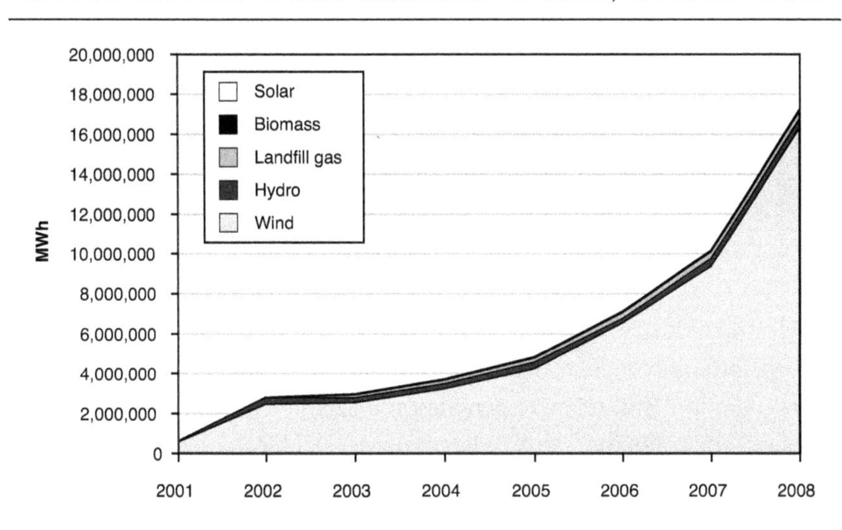

SOURCE: Prepared by the author with data from ERCOT's Renewable Energy Credit Program Information website (http://www.texasrenewables.com).

Energy-Efficiency Resource Standards. During the eightieth session of the Texas legislature in 2007, the Texas CHP Initiative recommended the creation of a energy efficiency resource standard for CHP modeled on the successful renewable portfolio standard (RPS). The concept acquired a sponsor, but it did not pass. The new standard would have required an additional 5,000 MW of CHP by 2016[47] and would have placed all technologies that provide environmental benefits on a more equal footing.

Renewable-Energy Credits and Credits for Energy Efficiency. The renewable-energy credits (REC) trading program in Texas is operated in pursuit of the Texas legislature's goal for renewable energy.[48] As figure 6-4 suggests, the original goal of installing an additional 2,000 MW of generating capacity in Texas from renewable-energy technologies by 2009 was easily surpassed, and the wind power market has grown rapidly, reaching 2 percent of total generation in 2007. The success of wind power has caused the Texas legislature and the PUCT to turn their attention to renewable resources other than wind as they continue to emphasize energy efficiency.

The role of CHP in providing energy efficiency presents opportunities for its projects to become eligible for energy-efficiency credit programs, particularly to satisfy requirements for an energy-efficiency resource standard (EERS). Expansion of retail choices and the extension of incentives for renewable energy to energy efficiency should give rise to an increase in small-scale CHP units.

Conclusion

The electric industry began in the 1880s with generation units sized to meet each customer's particular needs for mechanical power and light. "Distributed generation" still refers to generation located close to a customer's end use of electricity. As the industry has evolved, DG has become one of many competitive energy service alternatives for customers. The ability to sell excess power to utilities in the 1980s changed the economics of onsite power and resulted in rapid growth of combined heat and power. The utility purchase of economical, reliable nonutility power changed the electric industry forever.

Choosing among competitive energy services allows customers to specify the level of reliability they need, the quality of power suitable for their businesses, the price risk they wish to bear, and any other attributes of service they want to define. Customers who wish to capture and use waste heat on their premises can specify combined heat and power applications. Others may find competitive opportunities in wholesale markets through the sale of excess power from DG units, while still others may purchase some power or provide ancillary services to the transmission network. DG enhances the flexibility and energy security of the network, allows increased efficiency in the use of fossil fuels, and allows renewable resources and waste products to be put to use.

Since electric utilities given free rein as service providers might restrict DG opportunities to customers or competitive service providers, DG provides an excellent example of a competitive energy service that requires regulatory action to enable competitive markets and keep past anticompetitive practices in check. The Texas legislature and the PUCT established a platform for investment by DG customers and energy service providers, along with risk-taking and innovation. Changes in the production and use of electricity put past practices in jeopardy and increased uncertainty for utilities. As new roles were defined, electric utilities began to understand their new functions as market enablers and facilitators.

Understanding distributed generation as a competitive energy service should help the reader to understand that a robust retail electricity market involves technologies and services beyond the repackaging of power. A decade ago, regulated "plain vanilla" electric service was provided according to an approved tariff sheet. Today, in contrast, several DG options illustrate the multiplicity of products and services that customers might desire and receive through retail choice.

A maturing market for DG products and services will require the fuller participation of retail energy providers in Texas and the development of standard packages of services that can be marketed, financed, and constructed to meet the needs of customers in various market segments. Greater penetration of DG will increase the pressure for further industry reform, including, for example, pressure to link small generating units into smaller networks and increased need for local power transfers and transactions (sales to neighbors). Pressures for increased energy efficiency will also grow as energy costs rise and as carbon regulation internalizes external environmental costs.

We reject the notion that DG is merely a set of successful technologies in Texas. The value of DG arises from the ability of service providers to match a particular mix of technologies and services to a particular segment of the market. DG is one of a myriad of competitive energy services now available in the Texas electricity market. Large-scale CHP helped to transform the bulk power market, and smaller DG units located on the premises of commercial, institutional, and small industrial customers will help to transform the retail electricity market in the coming decade.

7

Competitive Performance of the ERCOT Wholesale Market

Steven L. Puller

Electricity markets around the world have been restructured using a variety of market designs. Policymakers have employed different mechanisms to organize the trading of electricity, price the use of the transmission grid, signal the value of new investment, and mitigate the exercise of market power. Each of the design choices can have substantial impacts on the competitive performance of the wholesale market. This chapter provides a brief survey of the market design of the Electric Reliability Council of Texas and summarizes recent evidence on the competitive performance of ERCOT's wholesale market.

Mechanics of Trading in the ERCOT Wholesale Market

The design of ERCOT envisioned a market that relies on bilateral trading as the primary means to buy and sell power. The vast majority of wholesale electricity transactions in ERCOT are bilateral transactions among generators, load-serving entities, and power marketers. Approximately 94–96 percent of ERCOT load is served through the bilateral market. The remainder is served by "spot" trades that occur through a centralized balancing energy market.[1]

Bilateral trades, which are conducted in over-the-counter markets such as the Intercontinental Exchange (ICE), can range from one-day deals to

This paper is based partly on research conducted jointly with Ali Hortaçsu. I thank Andy Kleit for his very helpful comments.

long-term transactions. One day before production and consumption occur, market participants (called "qualified scheduling entities," or QSEs) submit schedules of electricity to inject or withdraw at specific locations on the transmission grid. The day-ahead schedules may reflect a QSE's contractual obligations, but it is not required to do so. The balancing energy market ensures that generation and load balance each other in real time. For example, if the weather is unexpectedly hot or a generating unit suffers a forced outage, the balancing market is used to offset any differences between scheduled generation and real-time load. It can, in addition, serve as a means to "reshuffle" generation from the day-ahead schedule so that more efficient units are used to serve load.

The balancing market is a centralized auction that occurs just minutes before production and consumption. ERCOT determines the (nearly perfectly inelastic) demand in each fifteen-minute interval, and QSEs that are responsible for scheduling generation submit hourly offers to increase ("INC") and decrease ("DEC") energy relative to their day-ahead schedules. ERCOT clears the balancing market by intersecting the hourly aggregate bid function with the fifteen-minute demand function.[2] A generator called to INC is *paid* the market clearing price for all INC sales, while a generator called to DEC *pays* the market clearing price for the quantity of output reduced relative to the day-ahead schedule. In the theoretical literature on auction design, this approach to clearing the market is sometimes called a "uniform-price auction."

The price signals from the bilateral and balancing markets can serve two important purposes. First, the balancing price can create signals to facilitate least-cost dispatch. Because the day-ahead scheduled dispatch for each generator does not occur through a centralized market mechanism, there is no guarantee that the aggregate day-ahead schedule will approximate least-cost production. During an INC interval, the balancing price can signal the marginal cost of generation of the last unit of production. The balancing market can also send signals to replace less efficient units in the current schedule with more efficient units that have not been scheduled to produce. Generators' responses to these signals improve the efficiency of the real-time dispatch. Similarly, the balancing price during a DEC interval can signal which generation units should decrease production when load is less than the day-ahead scheduled generation. Thus, a critical condition for facilitating the lowest-cost dispatch of generation is that market-clearing balancing prices in both INC and DEC intervals equal (approximately) the competitive price of power.

Second, balancing and bilateral prices signal the value of new generation capacity. These spot and forward prices would be expected to converge if perfect arbitrage were to hold. In fact, forward prices reported in megawatt daily and balancing prices do not appear to diverge substantially.[3] ERCOT chose to pursue an "energy-only" model in which investment signals derive from the wholesale energy market rather than a market for "capacity."[4]

The efficiency of the signals from the bilateral and balancing markets is dependent on how close the market is to being competitive. In a competitive market, the market price is equal to the marginal cost of the highest-cost unit required to generate to serve load efficiently. In a centralized auction, this outcome would obtain if each generator submitted a bid schedule that corresponded to the marginal cost of generation.[5] Each unit with marginal cost below the perfectly competitive price would operate, while those with marginal cost above the competitive price would not. Any calculation of this perfectly competitive benchmark needs to account for constraints imposed by unit commitment (for example, startup costs) and outages.[6]

The market design in ERCOT differs from that of many other markets in the United States. In some markets, all generation and load are bid through a single auction. Dispatch is determined based upon these centralized bids and adjusted to reflect transmission congestion.[7] Such bid-based, day-ahead markets are used in the Pennsylvania–New Jersey–Maryland Interconnection (PJM) and by the New England ISO and the New York ISO and were used in the California market prior to 2001. Markets also differ in the means used to clear congestion, with some opting for locational prices that vary by zone and others for prices that vary by node.

Since its inception, ERCOT has used a zonal model to price congestion. In a zonal model, participants may wish to hedge against locational price differences. For example, a generator may wish to hedge price differences between the zone in which it sells its generated electricity and the zone in which it is obligated to buy power for its customers. ERCOT holds monthly and annual auctions for transmission congestion rights (TCRs), which essentially pay the price difference between two zones.[8] Zonal pricing does not, however, allow for highly differentiated signals as to the value of generation at specific locations. The goal of the new nodal pricing system to be introduced in 2009 is to use market-based signals to relieve transmission congestion in an efficient manner and to provide better incentives to build new generation at locations on the grid that best relieve congestion and provide reliability.

Trends in Generation Capacity and Load: Implications for Market Performance

The competitiveness of a market will depend upon load, the generation capacity and ownership of that capacity, and market rules. ERCOT load has grown continually, with peak load rising from nearly 57 GW in 2002 to 63 GW in 2006. Figure 7-1 shows the load duration curves in the hours with the top 5 percent of load.

A primary driver of wholesale prices is the price of fuel, particularly the price of natural gas. Generation in ERCOT is fueled by a mixture of natural gas, coal (lignite), nuclear, wind, and hydroelectric sources. The primary fuel is natural gas, with 76 percent of 2006-installed generation capacity being gas-fired steam and combined-cycle gas turbines. Because nuclear and coal units have lower variable costs, and market price is set by the marginal generating unit, the prevalence of gas-fired units makes them likely to set the market clearing price. As a result, natural gas prices are primary drivers of ERCOT electricity prices. This dependence upon gas prices is reflected in figure 7-2, which shows that prices in the balancing market were highest in 2005, when natural gas prices were relatively high.

Market designers must ensure that rules are in place to provide proper incentives for adequate generation resources, while also mitigating market power. Planning reserve margins have declined in recent years and stood at 16.5 percent in 2006.[9] If margins continue to fall, this trend will present a dilemma for market designers. On the one hand, market rules should ensure that proper incentives exist to signal the value of new investment in generation resources. On the other, a smaller reserve margin is likely to increase the potential market power for large generation owners.

It is inherently difficult for market designers and market monitors to distinguish scarcity rents from market power. To see this, consider the following two scenarios:

FIGURE 7-1

ERCOT LOAD DURATION CURVES, TOP 5 PERCENT OF HOURS, 2002–6

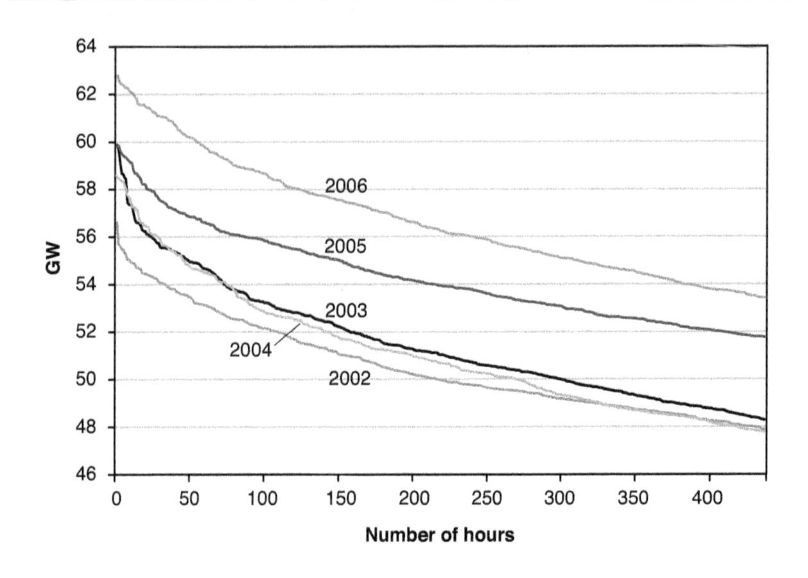

FREQUENCY OF HIGH DEMAND HOURS

Year	>58W	>56GW	>54GW
2002	0	2	34
2003	8	25	74
2004	8	29	76
2005	22	87	213
2006	124	239	381

SOURCE: Potomac Economics, *2006 State of the Market Report* (August 2007), Figure 46.

Scenario 1: When load is near capacity, large generators possess and exercise market power by bidding prices substantially above marginal cost.

Scenario 2: When load is near capacity, generators utilize all available capacity at prices close to marginal cost. Because expensive peaking units are called, the balancing prices are high.

FIGURE 7-2
ERCOT BALANCING PRICE DURATION CURVES, 2002–6

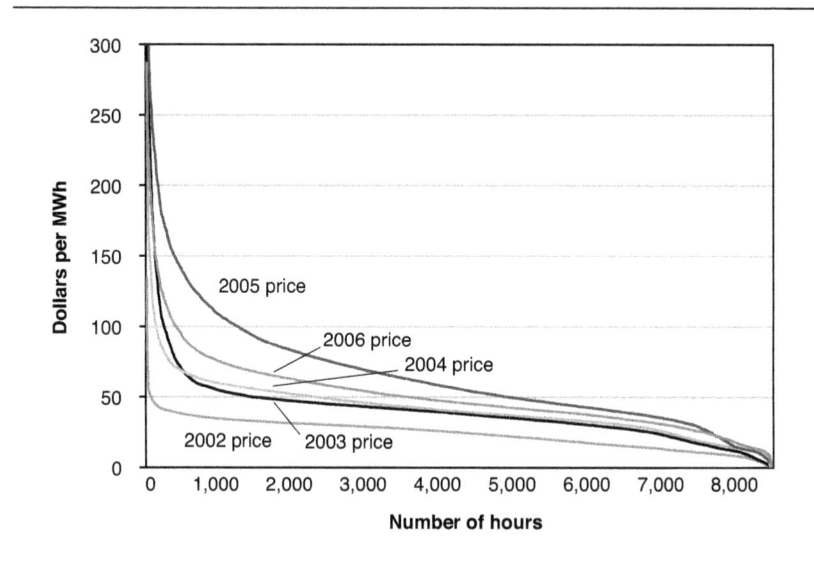

FREQUENCY OF HIGH PRICED HOURS

Year	>$300	>$200	>$100	>$50
2002	7	12	20	73
2003	22	80	254	1,473
2004	7	32	146	2,368
2005	45	175	1,285	5,047
2006	13	62	416	3,661

SOURCE: Potomac Economics, *2006 State of the Market Report* (August 2007), Figure 4.

These two scenarios call for different regulatory responses. Scenario 1 calls for market designers and monitors to mitigate market power because it generates production inefficiencies and high prices. Scenario 2 calls for the design of institutions to signal scarcity rents of new generation and thereby induce entry. Prices should not be mitigated in scenario 2 because high prices are signaling the value of new investment. These two scenarios may be observationally equivalent to the analyst. And even if economic analysis can distinguish between the two, politics may not. As one policymaker has noted, high prices are a political problem, whether caused by scarcity rents or market power.

This hypothetical regulatory dilemma illustrates that market power and resource adequacy are interrelated, and policies to address the issues should be coordinated. The Public Utility Commission of Texas (PUCT) recognized this relationship and designed rules regarding market power and resource adequacy together. In principle, the rules should allow price increases that reflect when generation capacity is scarce, but prevent generators that possess market power from exercising it. It remains to be seen how they will work in practice.[10]

Assessing the Competitiveness of the Wholesale Market

To evaluate the competitiveness of a wholesale electricity market, we must establish a benchmark against which to compare the actual market outcome. The ideal outcome is to achieve a dispatch that yields least-cost production, given constraints imposed by transmission congestion, ramping, and security constraints. In this setting, price would reflect the marginal cost of generation plus scarcity rents on new capacity. Because they have minimal demand response, electricity markets have always been plagued by the difficulty of allowing markets to determine scarcity rents. This problem has led to the administratively imposed value of lost load (VOLL), the creation of capacity markets, and market designs using energy-only approaches.

A market may still be "workably" competitive, even if it fails to achieve the benchmark of least-cost dispatch with prices that reflect the marginal generation costs plus scarcity rents. The relevant policy question is whether a given market design is better than any other. It is very unlikely that any actual market or regulatory mechanism could achieve a perfectly competitive benchmark. For example, nonmarket methods of procurement, including "old-style," rate-of-return regulation, have clearly generated inefficiencies. A bid-based market in which generators exercise substantial market power can also yield substantial inefficiencies. The goal of market design should be to achieve dispatch efficiency and prices that bring the system closer to the competitive benchmark. Incremental changes in design may provide the means to improve efficiency.

Ideally, an analysis of market performance would solve the problem of optimal dispatch and compare the optimal outcome to actual market dispatch and prices in both bilateral and balancing markets. Such "competitive benchmarking" approaches, which have been used to analyze market efficiency in both California and PJM,[11] are difficult to apply to the ERCOT market because the vast majority of transactions are bilateral, and data are not available on bilateral trades. Nevertheless, some assessment can be made of ERCOT's wholesale market performance using data from the balancing market.

To motivate the analysis, I provide a simple example to illustrate how a market with bilateral transactions and an imbalance market could yield an efficient outcome. Suppose that generating units vary in marginal cost ranging from $30 to $120. Suppose also that if all units are stacked up from lowest to highest marginal cost, system load will require those with marginal cost up to $80 to generate and those with marginal cost above $80 not to generate.[12] Now consider two QSEs—A and B. QSE A owns an (expensive) $100 unit and signs a large bilateral deal, and therefore schedules in the unit day-ahead. QSE B owns a (cheap) $60 unit and only schedules in one-half of the unit's available capacity. Efficient dispatch will require the $100 unit to DEC and the $60 unit to INC in the balancing market. This adjustment to the day-ahead schedule will occur if each generation owner bids into the balancing market at a price equal to the marginal cost. If, however, the owner of the $100 unit submits a (low) $40 DEC bid, and the owner of the $60 unit submits a (high) $120 INC bid, this reshuffling of the dispatch will not occur.

What could cause generators to fail to submit bids that would lead to the efficient reshuffling of generation? One possibility is that the QSEs have market power. In particular, if QSE B is a net seller of energy and therefore has market power on the INC side, it will submit bids at prices that exceed the marginal cost of generation. Similarly, if QSE A is a net buyer and therefore has market power on the DEC side, it will submit bids at prices below marginal cost.

Other possible causes exist, however, for failing to submit marginal cost bids into the balancing market. Suppose a QSE has small stakes in the balancing market and simply chooses not to participate in the auction process. This, too, would lead to inefficient dispatch. Below, I show evidence that many small market participants in ERCOT fail to submit balancing bids that allow the reshuffling of generation to occur. This evidence suggests that market monitors should consider a market design that encourages *all firms* (not only those with potential market power) to submit bids that minimize the cost of dispatch.

In a recent study by Hortaçsu and Puller, we tested the bidding behavior of each QSE into ERCOT's balancing market from September 2001 to January 2003, using hourly data on balancing bids, marginal cost of each generating unit, and the QSE's day-ahead schedule.[13] These data allowed us to compare the generator's actual bidding behavior to two benchmarks: perfectly competitive bidding, and bidding to exercise unilateral market power.

Under competitive bidding into a balancing market, a generation owner takes the day-ahead schedule as given and submits bids equal to the marginal cost of generation. On the INC side, the generator bids in any unloaded capacity of online units at marginal cost, up to the capacity of the unit. On the DEC side, the generator bids to reduce output at the marginal savings of lower output.

Under unilateral market power, the generation owner maximizes profit by choosing bids where marginal revenue of sales equals the marginal cost of production. Generators will, therefore, bid at a price above marginal cost to sell output larger than their forward contract position (that is, when they are net sellers). Similarly, generators will bid at a price lower than marginal cost to sell output smaller than their contract position (that is, when they are net buyers). Thus, if firms exercise unilateral market power, bid curves will be above the marginal cost for quantities greater than the contract position and below the marginal cost for quantities less than the contract position.

The logic of the empirical tests can be understood using figure 7-3. The marginal cost for firm i of supplying balancing energy is given by $MC_i(q)$. The firm's forward position is QC_i—it still "owes" its bilateral counterparties QC_i MW at the contract price. Thus, the firm is a net buyer until balancing sales are at least as large as QC_i, but it will be a net seller if its balancing sales are greater than QC_i. Suppose the firm submitted a bid function of $S_i^o(p)$ in the balancing market. The residual demand function $RD(p)$ is the total ERCOT-wide balancing demand minus the aggregate balancing bids by all rival generators. Then the market clearing price and the quantity of balancing power supplied is given by the intersection of the bid function $S_i^o(p)$

FIGURE 7-3
ANALYSIS OF ERCOT BALANCING BIDS

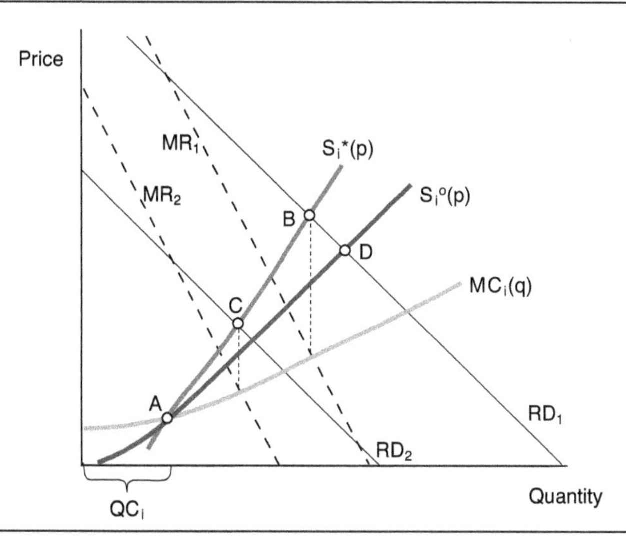

SOURCE: Author's illustration.

and the residual demand *RD(p)*. Given a particular realization of residual demand, RD_1, this firm would sell to the balancing market the quantity at the price given by point *D*. The profits can be calculated as the profits from meeting its contract position, plus the profits from its balancing sales.

Could the firm improve upon these profits? Economic theory teaches that profits are maximized by choosing the quantity and price where marginal revenue of sales equals the marginal cost of production. From the residual demand RD_1, the firm could calculate the marginal revenue function, MR_1, and select point B to maximize profits. Suppose instead that residual demand is smaller and given by RD_2. The profit-maximizing (price, quantity) point is given by point *C*. For any possible residual demand function, the firm can find the optimal point, and this set of points traces out the profit-maximizing bid function.[14] The optimal bid schedule is given by $S_i^*(p)$ and represents the exercise of unilateral market power. Our empirical test is to compare the actual bids that the firm submitted to two benchmarks: perfectly competitive bidding given by $MC_i(q)$, and the optimal exercise of unilateral market power given by $S_i^*(p)$.

Hortaçsu and Puller's analysis of bidding behavior focused on only a subset of time intervals when this methodological approach was most likely to be valid. In particular, we excluded intervals with substantial congestion and focused on one in which ramping constraints were unlikely to affect dispatch substantially.[15] Because of this limitation, we cannot extrapolate our results to other intervals.

What might one expect to discover when comparing actual bids to marginal cost? One might expect large firms with potential market power to exercise that power and submit INC or DEC bids substantially above or below marginal cost, respectively. Small firms might be expected to have little potential market power and to submit bids very close to marginal cost. This would suggest that market monitors should focus their attention on designing market institutions that might mitigate potential market power by the large generators; small generators are unlikely to contribute to inefficiency.

In fact, we arrived at a very surprising result. Indeed, the larger generators did appear to exercise market power. The smaller generators, however, generally did not submit bids close to marginal cost. In fact, they sometimes deviated from marginal cost-bidding more than the large generators did. This outcome is illustrated in figure 7-4, which shows bid functions for a large generator, Reliant, and a small generator, Guadalupe. Reliant had a contract position of approximately 500 MW this interval, giving it incentives to bid above or below marginal cost for balancing quantities greater or less, respectively, than 500. In fact, Reliant's actual bid closely approximated the optimal exercise of unilateral market power, differing rather substantially from marginal cost. This result was not surprising from the perspective of economic theory; one would expect firms to bid in a manner to maximize profits.

The bidding behavior by many small generators was unexpected, however. Guadalupe had a contract position of 0 MW and, because it was a small generator, only small incentives to bid above or below marginal cost for quantities greater or less, respectively, than 0. Even so, Guadalupe submitted bids that were substantially greater than marginal cost on the INC side and substantially less than marginal cost on the DEC side. Despite having generating units that could INC or DEC at a marginal cost of approximately $28, Guadalupe was unwilling to INC until the price rose to about $35, and unwilling to DEC until it fell to about $11. These bids were not

FIGURE 7-4
EXAMPLE OF BID FUNCTIONS

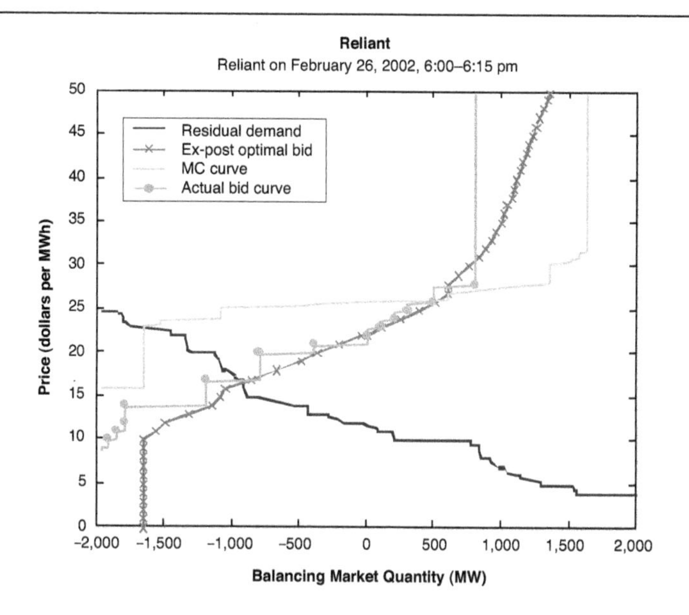

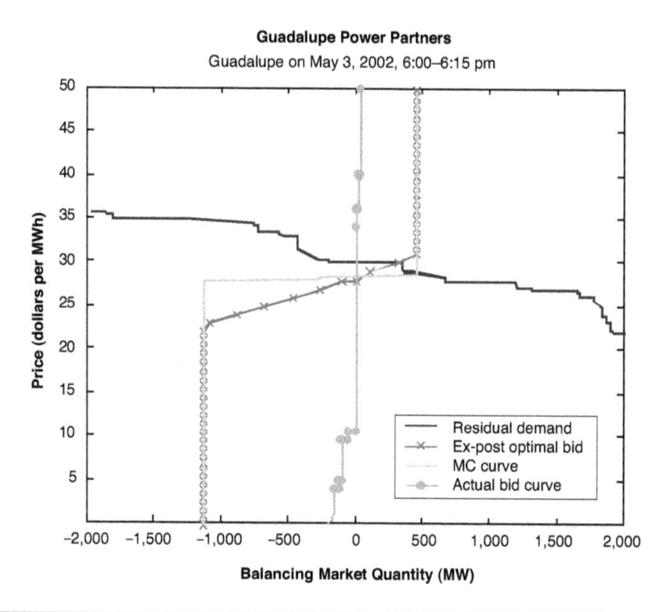

SOURCE: Hortaçsu and Puller, "Understanding Strategic Bidding in Multi-Unit Auctions" (2008).

consistent with the exercise of unilateral market power (shown by the "ex-post optimal bid" function). Rather, INC and DEC bids were much higher or lower, respectively, than bids to maximize unilateral profits. This suggests that bids significantly different from marginal cost are not intended as a means to exercise market power; generators exhibiting this behavior are sacrificing profits by bidding so high in the INC market or so low in the DEC market that they "price themselves out of the market." This behavior may be intended to avoid being called upon to change production from day-ahead schedules. Many small sellers in the analysis showed similar bidding patterns.

Hortaçsu and Puller explored a variety of possible explanations for these "excessively steep" bid functions by the small generators. We found little support for several explanations: unmeasured costs of adjusting generation, the possibility of transmission congestion, a dynamic pricing game, or characteristics of the generation technology. Instead, we found that one strong predictor of this behavior was having low stakes in the balancing market. This result suggests that small firms' steep bid schedules do not reflect attempts to exercise market power. Rather, small firms may not have sufficiently large dollar stakes to make it worth the time and cost of developing bidding strategies to participate in the balancing market. This, in turn, suggests that market designers should account for the participation costs of competing in complex, strategic environments when designing market mechanisms.

To illustrate this phenomenon, we studied all intervals for all generators in our sample of the early years of the market. We measured performance as the profits earned relative to a benchmark of not participating in the market. Generators that submitted vertical bid functions (that is, they bid to supply 0 MW at all prices) had a performance metric of 0, while generators optimally exercising market power had a performance metric of 1. From the sample figures, Reliant would be expected to have a metric close to 1, and Guadalupe closer to 0. Figure 7-5 plots this metric of performance against the quantity of sales the firm would make under optimal bidding. Firms with lower optimal sales had lower "stakes" in the balancing market. Figure 7-5 suggests that bidders with larger stakes in the market had a higher metric of performance. Larger firms were more likely to submit profit-maximizing bids and less likely to submit "excessively steep" bid functions. This strongly suggests that small generators do not appear to

FIGURE 7-5
MARKET PERFORMANCE VS. STAKES IN BALANCING MARKET

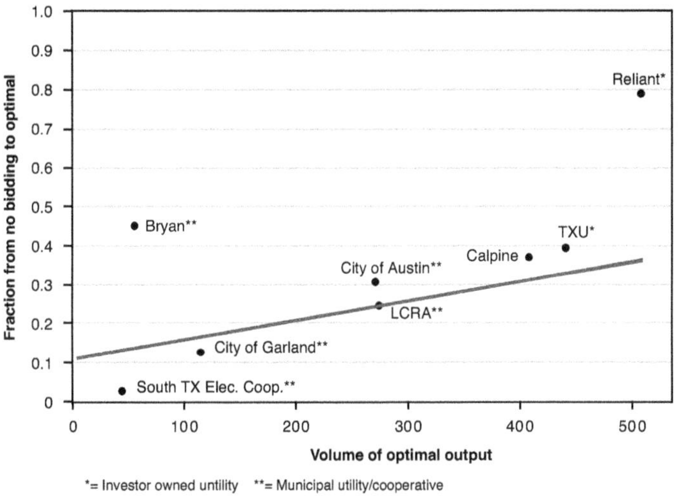

SOURCE: Hortaçsu and Puller, "Understanding Strategic Bidding in Multi-Unit Auctions" (2008).

select bids based upon sophisticated analyses of their strategic environment and, as a result, sacrifice profits in the balancing market.

Regardless of the cause of this behavior, the implications for efficiency are not trivial. As illustrated in the example above, generators bidding above marginal cost on the INC side may not be called upon to produce, despite having low-cost generators. Similarly, firms bidding below marginal cost of the DEC side may not be called upon to reduce production, even if they have high-cost plants operating.

Hortaçsu and Puller quantified the extent to which this behavior resulted in production costs higher than the least-cost dispatch. Least-cost dispatch occurred if each generator were to bid its marginal cost function. We estimated that the actual hourly dispatch costs of balancing energy were 27 percent higher than least-cost production. This suggested room for efficiency improvement if market designers could induce firms to submit bids closer to marginal cost.

What was the cause of this inefficiency? We decomposed it into two sources: the exercise of market power by the large generators, and the "excessively steep" bid functions by the small generators. This second source of inefficiency could arise from a variety of other sources, such as the fixed costs of establishing a sophisticated trading operation. We estimated that 81 percent of the total dispatch inefficiency arose from the bidding behavior of the smaller generators. Only 19 percent of the inefficiency was caused by the exercise of market power by the larger generators.

The results of Hortaçsu and Puller suggest that market designers should pay careful scrutiny to the bidding behavior of small generators. If this bidding behavior is systematic in all intervals of the day, and it continues today, then market designers should focus attention on encouraging greater participation by small generators. Inefficient dispatch can be caused by any bidding that substantially deviates from marginal cost (whether it is caused by market power, an unsophisticated bidding strategy, or anything else).

Any market rules that encourage greater participation by the smaller generators have two benefits. First, these generators will be called upon to generate when it is efficient to do so, and the dispatch costs will fall. Second, greater participation by the small generators will reduce market power by the large generators by making larger players face more elastic residual demand functions.

The Public Utility Commission of Texas rules implemented in 2006 address only some of the market inefficiencies. The rules properly identify market power as a possible concern in the balancing market and charge the independent market monitor to detect and prevent market manipulation. The PUCT rules explicitly exempt generation owners with less than 5 percent of installed capacity from having market power.[16] In addition, the practice of preventing market power appears to concern itself with market power on the INC side rather than on the DEC side (that is, on prices that are "too high" rather than "too low").

These findings suggest that focusing only on the bidding behavior by the largest generators during INC intervals can miss behavior that contributes to inefficient dispatch. First, the current market monitoring approach does not specifically address the lack of participation of small players in the balancing market. Of course, the motivation for small generators' steep bids is not likely market power. Nevertheless, it has exactly the same effect on market outcomes: Prices are higher than competitive prices during INC intervals, and the least-cost units are not generating.

Second, bids below marginal cost on the DEC side also contribute to inefficiency; high-cost units that would decrease output under least-cost dispatch are not called upon to do so. Market power on the DEC side has, however, received little regulatory scrutiny.

The results from Hortaçsu and Puller come with some qualifications. This analysis included only specific intervals (18:00–18:15) when ramping constraints were unlikely to bind, so the analysis was tractable. One might be concerned that behavior in other periods differed. Sioshansi and Oren conducted a similar analysis using all time periods of the day, however, and found very similar results.[17] In particular, their analysis suggested that some of the smaller generating firms were not rational strategic players in the balancing auction. These findings were consistent with there being participation costs to competing in centralized auctions.

In addition, we only analyzed the market's operation through January 2003; market design changes and monitoring by the PUCT and the independent market monitor may have altered generator behavior in recent years. Nevertheless, the results do provide strong evidence of a particular source of inefficiency that should attract scrutiny by market designers.

Conclusion

Any evaluation of the performance of a wholesale market should examine the actual market outcomes against a benchmark of efficient dispatch and competitive prices. The findings from such analyses should, however, be viewed with the recognition that any system of procurement will suffer from some form of inefficiency. It is for this reason that one frequently hears the term "workably competitive" applied as a criterion in evaluating wholesale markets. This chapter has identified some sources of inefficiency in the operation of ERCOT's market. While it is not necessarily possible to compare the size of the inefficiencies to those generated under another market design, these results suggest avenues for regulators and market monitors to pursue as they seek to improve the performance of ERCOT's wholesale market.

8

Retail Restructuring and Market Design in Texas

L. Lynne Kiesling

Retail competition became law in Texas on June 18, 1999, when Governor George W. Bush signed Senate Bill 7.[1] SB 7 provided for phased-in transition to full retail choice in the Electric Reliability Council of Texas (ERCOT) region by January 1, 2002, with a transition period, including pilot programs, commencing in 2001, and the institution of a retail-price transition mechanism for residential and small commercial customers called the "price to beat" (PTB). Although the implementation of SB 7 has faced challenges, most notably the political difficulty of designing and implementing retail restructuring and competition during a period of rising fuel input costs, it has successfully expanded choice for Texas consumers of all sizes, increased cost savings and production efficiency in the industry, and delivered improvements in environmental quality in Texas. The legislation is an example of careful, thoughtful, and practical design resulting in a resilient and adaptive institution.

This chapter describes the retail market design implemented in the state, highlighting the design features that have been the most important,

I am grateful to Eric Leslie for his valuable research assistance, to Andrew Kleit for his editing skills, and to Matthew Troxle at the Public Utility Commission of Texas for his guidance in finding relevant PUCT resources. Much of the background information in this chapter comes from the Scope of Competition Reports to the Texas legislature prepared by the PUCT from 1995 through 2007, and I am grateful for their thorough compilations of data and analyses. See, for example, Public Utility Commission of Texas, "Report on the Scope of Competition in Electricity Markets in Texas," Report to the 78th Texas Legislature (2003).

and the most controversial, in shaping market outcomes. It also analyzes some of these design features, particularly the fuel-cost adjustment for the PTB mechanism, in greater detail.

Prelude to Senate Bill 7

Texas's electricity restructuring began with policy changes that led to Senate Bill 7 in 1999. After SB 7's passage, the Public Utility Commission of Texas (PUCT) and ERCOT worked together to implement the technical, economic, and policy changes required to open the markets to full retail competition in 2007. These changes included a pilot project and a four-year transition period to allow for testing and for gradual adaptation to the new retail environment.

Several developments that preceded SB 7 created the foundation for retail competition. The Texas Public Utility Regulatory Act of 1995 (PURA95)[2] opened competition in wholesale power markets by allowing independent generators and power marketers to operate in Texas as of September 1995, as well as requiring open-access transmission service and pricing. ERCOT changed its mission and operations to become the first ISO in the United States in 1996. These changes fostered trade and competition in wholesale bulk power supply, although the many preexisting, long-term bilateral contracts limited their possible effect on end-use customers.

The competitive wholesale market did, however, attract significant generation investment to the state. In Texas, as in many other states, wholesale market liberalization preceded retail competition by several years. The primary argument for such a policy was, and continues to be, that retail competition cannot benefit consumers without robust, competitive wholesale markets. Under SB 7, areas of Texas outside of ERCOT cannot implement retail competition until the Public Utility Commission of Texas deems wholesale markets in their territories sufficiently competitive. This has not yet happened.

Before SB 7, few opportunities existed for retail competition, and those that were offered by the vertically integrated utilities generally targeted large commercial and industrial customers who could generate their own power. Self-generation and cogeneration make generation, transmission, and distribution potentially competitive by offering alternatives to vertically integrated utility service. In 1995, large industrial users generated 20 million MWh of electricity, which comprised more than 20 percent of industrial electricity consumption.[3] This potential competition also prompted utilities to offer discounted rates to large commercial and industrial consumers.

SB 7's Retail Market Design

The retail market design in Texas focused on several dimensions of making the transition from a vertically integrated, regulated utility to an unbundled industry with competitive generation and retail, using regulated transmission and distribution wires services. The market design focused both on structural rules leading to changes in the firms themselves and behavioral rules that shaped the incentives facing incumbents and new entrants.

Unbundling and Participation. SB 7 laid the foundation for opening the Texas market to rival retail electricity providers (REPs). REPs are parties that purchase power from generators and provide customer service to the final customer. SB 7 required incumbent utilities to unbundle the four components of the traditional vertically integrated supply chain: generation, transmission, distribution, and retail marketing. The PUCT staff summarized the economic logic supporting this policy in their pre-restructuring analysis:

> In the past, electricity has been assumed to be a "natural monopoly," where a single firm can provide service at a lower cost and higher effectiveness than multiple firms competing with each other. Today it is clear that at minimum the generation function can be fully competitive, rather than monopolized by a single firm. And if the integration between utility generation and distribution is broken, it is no longer necessary that electricity delivery—the actual retail power delivery and customer service function, as distinct from the function of building and operating the distribution wires network safely and reliably—be provided only by a single, monopoly firm.[4]

The form of the unbundling was largely functional, not structural; incumbents could retain ownership of unregulated retail affiliates and of generation capacity, but had to create separate functional units for power generation, power delivery (the transmission and distribution utility, or TDU), and retail sales.[5] They were also required to divest their generation assets in excess of 20 percent of their native service territory load. Municipal utilities and cooperatives were not required to participate, but they could choose to do so.

The Price-to-Beat Transition Mechanism. The "price-to-beat" mechanism included in SB 7 provided some retail rate stability by ensuring that a fixed, regulated rate would be available for small customers during the transition period.[6] Under it, an affiliated retail electric provider (AREP) could not charge its native residential and small commercial customers a rate lower than its PTB rate.[7] This price floor remained in effect until January 1, 2005, or until the AREP lost 40 percent of its residential and small commercial load, at which point the AREP could lower its prices. Between January 1, 2002, and December 31, 2006, the PTB would also act as a price ceiling, providing a default service contract for small customers in the AREP's native service territory. As the end of the transition period approached, the PUCT could evaluate whether a retail market was sufficiently competitive to allow the PTB to expire on January 1, 2007, as planned.

The PUCT's initial determination of the PTB rate for each of the five incumbent territories incorporated several components.[8] First were the mandatory charges that would be included in the TDU component of all retail rates: transmission and distribution charges, metering charges, other customer charges, the nuclear decommissioning charge, and the system benefit fund charge. Second, the PTB rate reflected the wholesale cost of purchasing electricity; and, third, it allowed for costs of retail sales and marketing. Finally, the PTB included "headroom," or a profit margin that would enable potential entrants to come into the market and compete against it. In each region, the PTB rate reflected a 6 percent discount on the regulated rate that had been in place on January 1, 1999. Figure 8-1 illustrates the composition of the PTB.

The PUCT established the initial PTB fuel factor in much the same way as it would have been established under regulation: by forecasting fuel

FIGURE 8-1
COMPONENTS OF THE PRICE-TO-BEAT TRANSITION MECHANISM

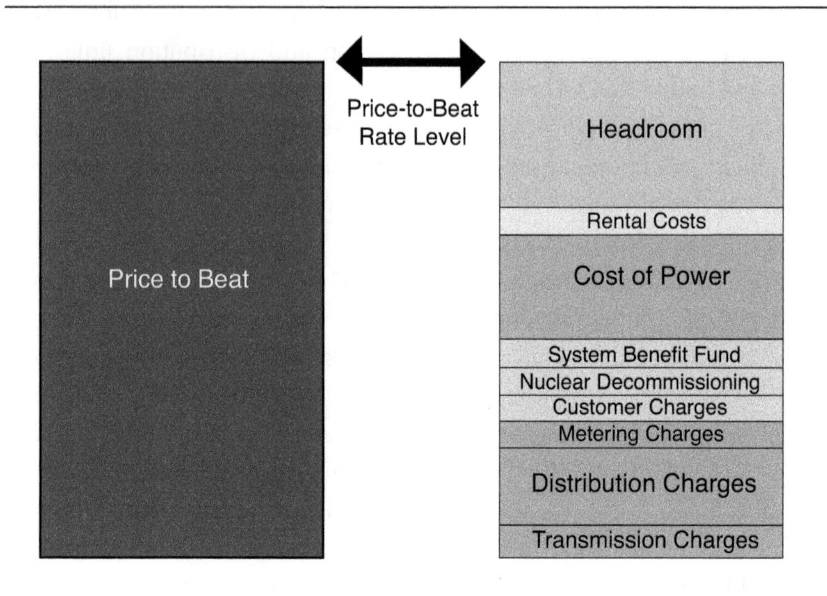

SOURCE: Public Utility Commission of Texas, "Report on the Scope of Competition in Electricity Markets in Texas," *Report to the 78th Texas Legislature* (2003), 20.

prices and taking into account the likely generation fuel mix. It recognized, however, that the fuel factor should be flexible enough to cope with the potential volatility of fuel (and, consequently, of electricity) prices. Having analyzed the market designs in other states, Texas legislators and regulators concluded that a flexible PTB would do a better job of promoting competition than a fixed standard-offer price. California's experience with a fixed standard-offer price and PG&E's bankruptcy filing as a consequence of rising wholesale electricity prices reinforced this design choice.[9]

The PUCT further found that for all Texas retailers, the price of natural gas was such a substantial determinant of wholesale electricity prices that the fuel factor adjustment should be based on changes in natural gas prices, as well as on total electricity purchases.[10] Under the PTB mechanism, an AREP could request a fuel factor adjustment if natural gas prices changed by at least 4 percent.[11]

The PTB provision for wholesale energy costs was controversial even at its inception because of this use of natural gas futures prices as a benchmark. Some were concerned that the benchmark would overstate changes in costs because not all of the electricity in Texas was generated using natural gas. If, for example, the price of natural gas increased but the price of coal did not, under the PTB the AREP could attain a fuel factor increase on all megawatt hours sold, not just those generated using natural gas. While true, that argument overlooked the fact that what matters for determining market prices is the marginal cost of the marginal generating unit, not the average cost over all generating units. (In addition, prices for all sources of energy generally move together.) For that reason, the natural gas price benchmark was appropriate.

Provider of Last Resort. SB 7 also required the PUCT to designate provider of last resort (POLR) REPs for each area to serve residential customers who did not actively choose a REP, or whose REP had left the market. Each POLR would offer a standard retail service package at a fixed rate[12] and serve the customers of REPs that exited the market or otherwise failed to serve them. This included nonpaying customers whose service was discontinued; the POLR was the only retailer that could disconnect them.

After some criticism and difficulty in attracting providers of last resort, the PUCT revised the rules to remove nonpaying customers from the POLR category as of September 2002. Under the new rule, AREPs were required to serve nonpaying customers at the PTB, and any REP could disconnect customers if the PUCT found that retail consumer protection provisions existed in the region. POLRs could adjust the fuel portion of their rate monthly to reflect wholesale electricity price changes, protecting their customers from short-term price volatility while protecting the providers from the risk associated with a fixed retail price.

Renewable Energy. The issues of renewable energy and the environmental impact of increasing electricity demand also drew the attention of legislators and regulators. SB 7 contained a renewable portfolio standard (RPS), which was designed and implemented in late 1999.[13] The legislation required the construction of 400 MW of renewable energy resources by 2003 and 850 MW by January 2005, scaling up to installations of 2,000 MW in total by

2009. The federal production tax credit of 1.5 cents per kWh, enacted in the federal Energy Policy Act of 1992, also helped speed up efforts to meet this target. Expired and extended three times since 1999, the tax credit has, in conjunction with the RPS, induced investment in new renewable energy resources, particularly wind power. SB 7 also contained an emissions-reduction mandate of 50 percent for nitrogen oxides and 25 percent for sulfur dioxide from electricity generation.

In 2001, the share of renewable energy in Texas' portfolio mix was 1 percent, which was relatively low. Before restructuring, most renewable energy used in Texas was hydroelectric; since 2001, the renewable portfolio standard has contributed to the development of substantial wind energy resources in West Texas.

Consumer Education, Protection, and the System Benefit Fund. Retail restructuring had the potential to create unanticipated and hitherto unknown outcomes for consumers, particularly for residential customers who had not previously had opportunities for retail choice. Although increased choice and efficiency and the ensuing cost reduction were likely to make even vulnerable consumers better off, there were no guarantees. With residential consumers lacking experience in shopping for power service, the legislature had concerns about retailer transparency and business practices.

These concerns led to provisions in SB 7 for consumer education and protection, with the objective of ensuring that consumers, including low-income and non-English-speaking consumers, were sufficiently knowledgeable about retail competition and choice to make informed decisions. The PUCT's Texas Electric Choice campaign began in February 2001. Phase 1 focused on awareness in advance of the pilot project and transition period; phase 2 emphasized how to choose an electric provider. The education campaign had six messages to communicate:

- Electric choice was working for Texas.
- Customers should become informed.
- The PUCT would continue to protect customer rights.
- The transition to a competitive electric market would take time.

- Customers should learn how to shop for electricity.
- Customers should compare offers from retail electric providers and explore their options.[14]

The PUCT implemented the Texas Electric Choice campaign in several ways: through the use of paid advertising, a toll-free billing call center, websites in English and Spanish,[15] newspaper inserts, and other educational literature.

SB 7 also coupled the phased implementation of retail competition with strong provisions for consumer protection, including the Consumer Protection Division, the Office of Consumer Protection, and the system benefit fund. These provisions built on the establishment of the PUCT's Consumer Protection Division in 1997 and the Office of Consumer Protection, which is a Texas state agency that investigates consumer protection issues generally, in 1998. The PUCT's Consumer Protection Division receives and responds to consumer complaints about electricity and telecom service; most electricity complaints are related to billing.

The provisions in SB 7 for creating and funding a system benefit fund (SBF) were intended to assist low-income consumers with budget difficulties. Although each utility established its own assistance plan for low-income customers under regulation, such a plan was not likely to be feasible under retail competition. The SBF included in the wires charge a mandatory customer charge per kilowatt-hour to fund four revenue streams:

- Rate discounts for low-income customers;
- PUCT customer education activities (described above);
- the building of weatherization funding targeted at low-income customers; and
- lost property tax revenue.

The issue of lost property tax revenue arose because of concerns about possible decreases in utility asset values associated with restructuring. Property taxes on those assets were needed to fund school districts and other state functions, and changes in asset values in the local tax base might create revenue shortfalls.[16]

During the implementation of SB 7, the administration of the SBF focused on the discounts to low-income consumers, which the PUCT implemented as the Low-Income Telephone and Electricity Utilities Program (LITE-UP). LITE-UP would provide a 10–20 percent discount based on income qualification during four summer months (June–September).

The 2001 Pilot Project. As part of the transition to retail competition, seven utilities (including two outside of ERCOT that had not, to date, adopted retail competition) participated in the Texas Electricity Choice Pilot Project. Starting June 1, 2001, 5 percent of the load in each utility's service territory could choose their retail provider. The pilot project served three important functions by providing an opportunity to test computer and communication systems, informing and educating consumers, and exploring the potential for retail competition at the regional level.

By the end of the pilot on December 31, 2001, over 115,000 customers had participated. Of them, 90 percent were residential, 9 percent nonresidential with peak demand less than 1 MW, and 1 percent nonresidential with peak demand greater than 1 MW. One of the most useful roles of the pilot project was to serve as a venue for identifying and resolving technical problems before implementing widespread retail choice in 2002. Most of these problems involved communicating data among customers, REPs, TDUs, and ERCOT in the course of processing switching transactions. While most of these technical issues were addressed before the end of the pilot, not all were resolved; however, the PUCT decided not to delay the market start date. At the same time the pilot project was under way, the commission also heard administrative cases to set the initial PTB for each region in fall 2001.

Transition Period, 2002–6. The transition period began on January 1, 2002. Large commercial and industrial customers (with peak demand greater than 1 MW) were already prepared for their rapid transition, as they had neither a PTB price transition mechanism nor designated POLRs with which to contend. All residential and small commercial and industrial customers who had not chosen a REP were transferred from the utility to the AREP on January 1, 2002, at the PTB rate.

Although the PTB rate represented a 6 percent discount on the regulated rate, new REPs were entering the market in all regions and could compete against the AREP's PTB. The fact that REPs were entering the market and providing competitive residential services suggests that they anticipated being able to compete against the discounted PTB rate, which was consistent with the objectives of the PTB and the inclusion of headroom in it. For the most part, prices were the basis for this retail competition, with little product differentiation other than some green-power contracts and the entry of Green Mountain Energy. One of the first, and most striking, manifestations was the cross-AREP competition that occurred; in particular, TXU (with a native territory around Dallas) and Centerpoint/Reliant (with a native territory around Houston) entered each other's native service territories early in 2002. By September 2002, twenty-five REPs had entered the Texas markets, and residential customers had between three and ten REPs from which to choose, depending on their locations.

This rivalry generated very large savings for consumers. Even those customers who stayed with their AREPs on the PTB saved $900 million in 2002 relative to their regulated 2001 rates—$225 million due to the mandatory 6 percent discount, and $675 million due to lower fuel prices that year. In the first nine months of the transition, 400,000 residential customers had switched.[17] Commercial and industrial customers also benefited from competition; they could choose from among nineteen REPs and paid lower rates in 2002 than in 2001.

By September 2004, more than 1 million retail customers had switched to a REP, and ERCOT had processed 1.5 million switch requests. REP entry grew, as did product differentiation; each region had from seven to twelve REPs offering from nine to fourteen products to residential customers. The years 2003 and 2004 were challenging for retail restructuring, though, because of the higher natural gas prices that led to increases in both wholesale electricity prices and the PTB. Between the commencement of full retail competition in January 2002 and September 2004, natural gas prices increased by 150 percent, and ERCOT's average wholesale electricity price doubled.[18] Between January 2003 and September 2004, AREPs filed sixteen requests for PTB fuel factor changes, leading to a 20–35 percent increase in the PTB.

The rapid rise in natural gas prices in late 2004 led to persistent high prices, exacerbated by a sharp spike in August and September 2005. The damage inflicted by hurricanes Katrina and Rita on natural gas facilities around the Gulf of Mexico caused this dramatic increase; as repairs were made and the economy absorbed the shock, natural gas prices fell in 2006 and remained relatively stable in 2007.

As a consequence of the high natural gas prices in 2005, AREPs requested PTB increases. PTB rates that had been 8–9 cents/kWh in 2002 rose to 12–13.5 cents/kWh in 2005.[19] As the prices fell in 2006, however, AREPs did not reduce their PTB rates.

By October 2006, seventeen REPs were serving 1.8 million residential customers—double the number from two years earlier and 34 percent of all residential customers—and offering thirty-five to forty-one products per region. Those who switched were saving 16–31 percent relative to the PTB, an indication of the potential benefits presented by competition in the wake of the AREP decisions not to reduce the PTB in 2006. In some regions, consumers could purchase premium green-power products at prices lower than the PTB.[20]

Investment in renewable energy resources also increased during the transition period, as the RPS designers had intended. By October 2002, 1,000 MW of renewable energy resources had been installed, achieving the interim goal more than a year early and then exceeding it. The installations increased to 1,187 in November 2004, 3,262 in November 2006, and 4,632 in November 2007.[21]

One obstacle to this progress was that most of the wind power installations were located in West Texas, and transmission constraints limited the ability to deliver that power to other parts of the state, thus resulting in transmission congestion costs.[22] The PUCT responded by amending the transmission-planning process in late 2006 to expedite transmission expansion between West Texas and the eastern part of the state. Rules were established for designating "competitive renewable energy zones," with an accelerated permitting process for transmission serving them.[23]

Early in the transition period, consumer confusion and complaints emerged over technical issues with respect to switching, customer moves in and out of premises, and metering and billing. By May 2002, however, ERCOT was processing 90 percent of switches correctly, and, by November 2002, all of the backdated switches were processed and the billing timelines returned to their historical benchmarks.[24]

Complaints to the PUCT's Consumer Protection Division also increased during the first few months of the transition period with regard to billing problems or complications in processing switching transactions. As REPs became more familiar with the new market, these complaints declined. Some consumer issues remain, however. For example, while REPs compete on per-kWh prices, the fixed charges on the final bill are not entirely transparent to the end-user.

Problems also persist concerning POLR service for high-credit-risk customers. One way that REPs address this credit risk is by asking customers to make a security deposit before they receive service. Paying a deposit, though, is beyond the budgets of some low-income consumers. One long-term approach being considered is prepaid service, but such a service cannot be implemented until the TDUs have installed advanced metering infrastructure (AMI). In some areas, such as the Oncor (formerly TXU) territory in the Dallas–Fort Worth area and the Centerpoint service territory in the Houston area, AMI was scheduled to be rolled out in phases to residential customers in 2009.

During the changes of the transition period, the consumer education campaign continued. In a survey in August 2002, over two-thirds of residential customers and almost 80 percent of commercial and industrial customers indicated that they were aware of restructuring. From 2000 to 2002, residential customers who believed they knew "a great deal" or "a moderate amount" about competition increased from 15.3 percent to 62 percent. More than half of residential customers viewed competition positively, and those who viewed it negatively generally did not see sufficient net benefits.[25] In 2004 and 2005, the Texas legislature reduced consumer education funding, hampering the PUCT's efforts to educate consumers. Unfortunately, the reduction coincided with the first substantial increases in natural gas prices and PTB fuel factors.

Performance of the PTB Fuel Factor Adjustment Analyzed

The price-to-beat concept played an important role in Texas's approach to electricity deregulation. Recall that the PTB was pegged at 6 percent below a static price at a fixed time, with allowance for increases in fuel costs. For

FIGURE 8-2

PRICE OF NATURAL GAS DURING ACTIVE PRICE-TO-BEAT MECHANISM

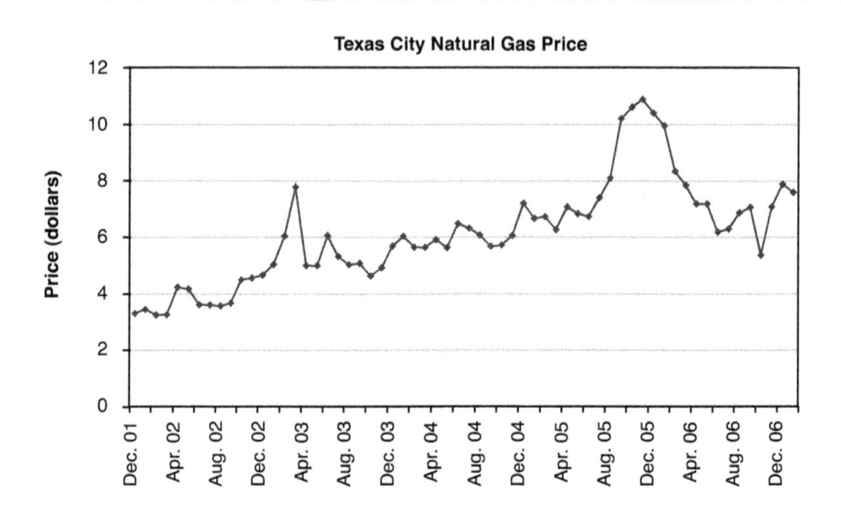

SOURCE: Energy Information Administration, U.S. Department of Energy, *Electric Power Annual 2006* (2007).

retail restructuring to succeed, the PTB had to operate as a mechanism to encourage competitive rivalry without causing financial hardship for the AREPs. Thus, a provision was included in SB 7 that allowed AREPs to request increases in their PTB levels up to twice per year.

As I have already mentioned, natural gas is the predominant fuel input for electricity production in Texas, and thus its price is one of the primary factors in determining the price to beat for Texas utilities. Many factors influence the price of natural gas, including weather, demographics, economic growth, fuel competition, storage levels, and exports. During the years 2002–6, natural gas prices ranged from $1.76 to $10.87 per thousand cubic feet, as shown in figure 8-2, and were especially volatile in the aftermath of Hurricane Katrina in 2005.

As the price of natural gas rose, the six AREPs exhibited similar behavior in their requests for price-to-beat increases. Every request for a price increase made by a utility was followed by a processing lag—often lasting a couple of months—before the increase was granted. It is difficult to tell whether price increases were made according to the natural gas prices prevailing at the time of request.

In May 2002, the AREPs requested their first price increases within days of each other. (Note the increase at that time in figure 8-2.) These initial requests corresponded approximately to a cumulative increase of 27 percent in the PTB since its inception in January 2002. These increases were approximately the same as the corresponding natural gas price increase, in percentage terms, at the time (26.8 percent). In August 2002, the AREPs were granted further increases averaging 20.5 percent. By March 2003, the utilities had submitted their second or third requests for fuel factor increases, by which time natural gas prices had risen by 136 percent since the PTB's inception. Even after the 2003 PTB increases, no AREP's PTB had risen by more than 66 percent. Thus, by March 2003, natural gas prices had risen twice as much as the PTB.

After that, however, the cumulative percentage increases in the PTB surpassed those in the natural gas price (see figure 8-3). As more generous increases in excess of the rate of movement in the natural gas price were granted to the utilities, the prices started to rise at a much faster rate, further compounding the issue.

As figure 8-4 shows for 2002–6, the greatest month-to-month increases in the utilities' price to beat corresponded to those in natural gas. For example, in February 2003, natural gas prices rose by more than 25 percent in a single month, which prompted PTB increases (of 36 percent, 23 percent, 14 percent, and 35 percent, respectively, for four of the AREPs). In November 2005, gas prices increased by 15.7 percent, followed by PTB fuel factor increases for three AREPs (of 23 percent, 15 percent, and 20 percent, respectively).

In 2004, these PTB fuel factor increases began to appear unjustified to many consumers. Local Texas newspapers consistently voiced consumers' discontent with the PTB system and, by extension, electricity restructuring in general, stemming from rising electricity prices that stayed high even after natural gas prices had begun falling. The price-floor aspect of the PTB angered consumers facing ever increasing electricity bills. The *San Antonio Express-News* commented that the "tumult" of deregulation began when the PUCT set the price to beat. The *Fort Worth Star-Telegram* stated, "A genuine competitive market for electricity production has not developed," and

FIGURE 8-3
CUMULATIVE PERCENTAGE CHANGES IN NATURAL GAS PRICE AND PRICE TO BEAT

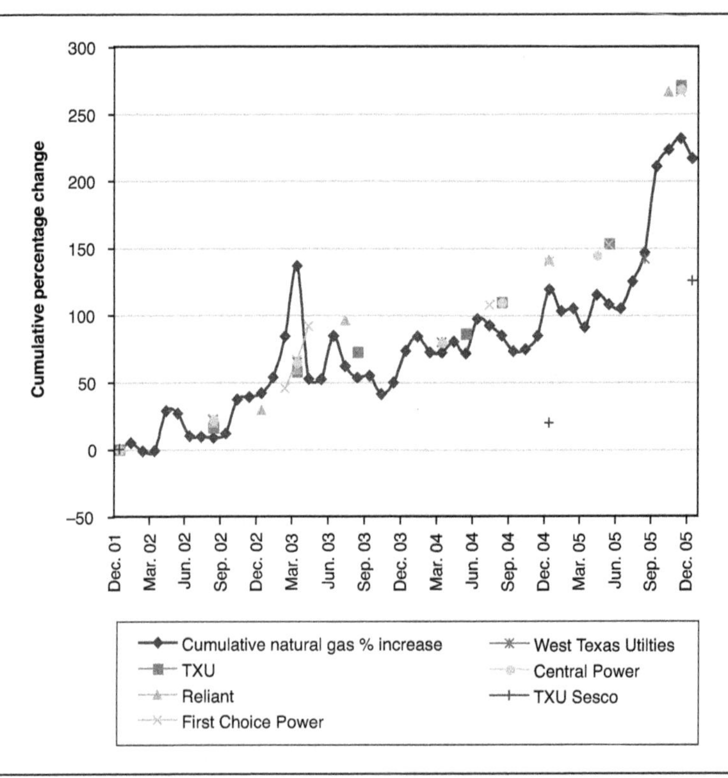

SOURCES: Energy Information Administration, *Electric Power Annual 2006*; Public Utility Commission of Texas, "Electric Power Industry Scope of Competition and Potential Stranded Investment Report, Volume II," Report to the 75th Texas Legislature (1997).

quoted candidates for Texas governor describing the price to beat as "an unmitigated disaster."

One commonly recommended solution was to give the PUCT the power to adjust the PTB downward. In February 2006, the *Fort Worth Star-Telegram* explained that consumer groups had been pushing the PUCT to rescind some of the previously granted price increases that were no longer necessary in the face of falling natural gas prices. Earlier, in April 2002, the Office of Public Utility Counsel had unsuccessfully taken the PUCT to court to establish an increased price to beat as unlawful. According to the *Dallas Morning*

FIGURE 8-4
MONTH-TO-MONTH PERCENTAGE CHANGES IN THE PRICE OF NATURAL GAS AND PRICE TO BEAT

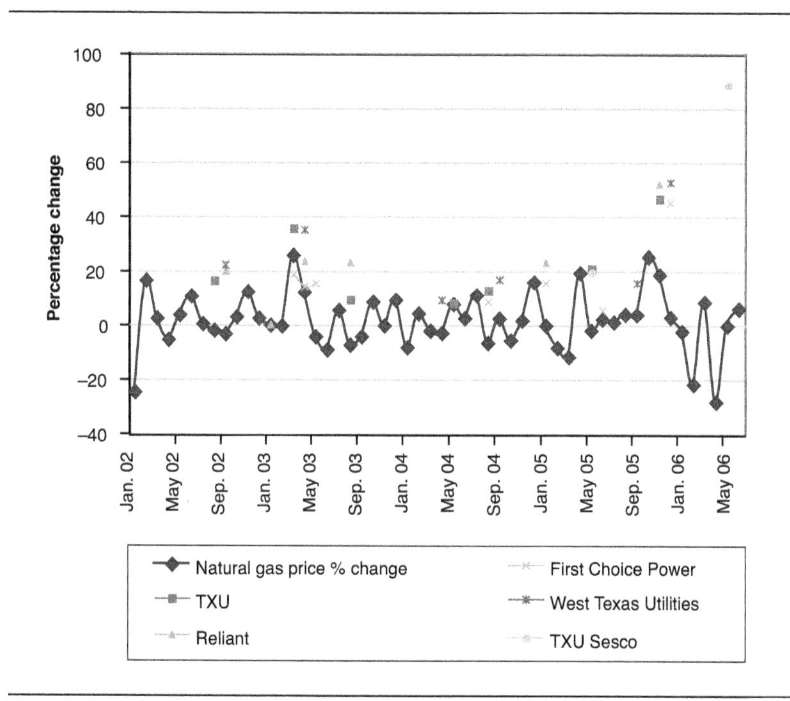

SOURCES: Energy Information Administration, *Electric Power Annual 2006*; Public Utility Commission of Texas, "Electric Power Industry Scope of Competition Report, vol. II" (1997).

News, one of the biggest concerns of consumer groups was that the PTB was set artificially high to induce customers to switch. Some media authors wrote in support of the PTB; however, the general media sentiment was that it should be revoked or revised.[26]

Despite the widespread discontent in the years following the introduction of the PTB, we should note that even at the point when natural gas prices were the highest, in late 2005 and early 2006, the price of natural gas quadrupled while the average PTB increased only by 90 percent. In part, AREPs could not pass on the entire short-term price increase because of retail rivalry. Also remarkable is that the AREPs chose not the decrease the PTB in 2006 to compete more directly on price with the REPs. One inference to draw is that they thought consumer inertia would be high enough to make it more profitable to retain the high PTB than to engage in price rivalry.

Current Snapshot: Outcomes and Consequences of Retail Competition

The PTB mechanism expired on January 1, 2007, opening the full Texas market to retail competition. How are consumers and producers faring now, compared to the regulated environment in 1999? Is there evidence that total surplus has increased? We should avoid committing Demsetz's "nirvana fallacy,"[27] and should evaluate this institutional change relative to feasible benchmarks rather than unattainable theoretical ones. We should base our evaluation on evolutionary criteria: How adaptable is this design to unknown and changing conditions?

Although the high natural gas prices and hurricanes in 2005 induced increases in both the PTB and the prices offered by REPs, the Texas retail market design adapted to those strains, and today it shows demonstrable evidence of success. Many competitors have entered and offered a variety of products, customer switching has been active, AREPs have changed their product offerings to retain customers, and new investment has occurred, with more planned. Nevertheless, challenges remain, including billing transparency, customer credit issues, and TDU installation and implementation of widespread AMI.

Texas customers can choose from a wide range of retailers and products, amounting to over ninety-five products from more than twenty-five providers in each region, and enough switching has occurred that REPs now provide the majority of energy sold (across all consumer classes) in the ERCOT region. In comparison both to the performance of a regulated market and that of retail restructuring in other states, the extent and variety of choice in Texas are impressive. Judging by entry, rivalry, and product differentiation, this institutional change has been successful. When performance is measured by number of customers instead of by energy consumption, 40 percent of residential customers have chosen a REP, which also indicates the success of retail competition in Texas under difficult circumstances.[28]

The most striking differences between current market conditions and regulated conditions are evident in both retail and wholesale prices. Recall that the last regulated rate in 2001 included a 6 percent discount on previous rates, and that since 2001 natural gas prices had tripled. Six years later, three of the five regions with retail competition had competitive offers available in the market that were lower than the last regulated rate—a reflection of the competitive pressure exerted by retail competition on wholesale markets as well. In 2006, fuel costs were lower than they were in 2005, and the competitiveness of ERCOT's markets led to the largest percentage decrease in wholesale electricity prices in the United States.[29]

Another change that has taken place in Texas since the passage of Senate Bill 7 has been the greater investment in, and use of, renewable energy. This increase is not entirely market-driven but is largely the consequence of the renewable portfolio standard provisions of SB 7 and federal tax credits for the construction of new renewable capacity. One market driver of renewables is the differentiation among products that has occurred with retail choice. Both REP entry (that is, Green Mountain Energy) and offerings of renewable products from REPs and AREPs enable customers to choose renewable power. Most REPs and AREPs offer at least one renewable energy product. Renewable capacity has increased by 390 percent since 1999 and is differentiated to include wind, biomass, landfill gas, and solar in addition to the hydroelectric capacity that existed before SB 7. And as the renewable portion of the fuel mix has doubled, Texas has seen a corresponding decrease in nitrogen oxide emissions (57 percent) and sulfur dioxide emissions (22 percent) from electricity-generating sources,[30] much of it due to the retirement of older, less efficient generation capacity and investment in more efficient, cleaner generation technology.

One of the hardest things to see when evaluating actual results is the counterfactual: What would have happened in Texas if restructuring had not occurred? This question is particularly pressing in light of the natural gas price increases between 1999 and 2007. Had the state still been under economic regulation, these fuel price increases would have raised regulated rates, because regulated utilities were entitled to fuel factor adjustments by law.[31] Thus, an "apples to apples" examination requires comparing actual retail prices with an estimate of what the regulated rates would have been.

In 2006, the PUCT estimated this counterfactual for the Centerpoint/Reliant and TXU service areas for the years 2002–5. They found that in the Centerpoint/Reliant area, the estimated regulated price would have been 18–26 percent higher than the average of the five lowest actual retail prices. In the TXU area, it would have been 11–18 percent higher.[32] This analysis suggests that the results from retail restructuring compare favorably to regulation, even without taking into account the benefits of more generation portfolio diversity, the retirement of older, less efficient generation, higher air quality, increased consumer protection, and the instrumental value of consumer choice and empowerment in and of itself.

We may also measure success in Texas by comparing the state with others that have nominally removed the price restrictions on retail competition. New York consumers have had less switching and fewer product options than Texas consumers. Residential retail choice in Illinois and Maryland has been stagnant because of the politicization of the dramatic move from a fixed-rate cap to no rate cap, combined with the extensive use of long-term wholesale procurement contracts for the incumbent's default service product. By avoiding these two pitfalls, Texas has succeeded in creating a competitive retail environment that benefits consumers and retailers while also protecting low-income consumers and promoting both innovation and environmental quality.

Conclusion

The opening of retail markets in Texas has generally been a success in a trying environment of rising input costs. The openness and the stakeholder focus of the restructuring process, the inclusion of consumer education and protection, the transition period, and the price-to-beat mechanism with an adjustable fuel factor have adapted to changes in underlying fuel costs better than has happened in states with long-term retail rate freezes. In the face of challenging market conditions in 2005, with natural gas price increases and two hurricanes, Texas's integrated wholesale and retail markets worked as they were supposed to: They reflected cost increases and decreases and gave consumers opportunities to adjust their behavior in ways that are impossible under retail rate regulation. Retail choice enabled consumers to be more adaptive, leading to a more resilient economic system, and the conditions of 2005 tested and demonstrated that resiliency in Texas's competitive retail electricity market.

One of the most remarkable aspects of Texas's electricity restructuring in general, and the retail component of it in particular, is that policymakers recognized that the industry was not a natural monopoly anymore, and that competition would serve the public interest. They took action to revise their policy mission to reflect the role that competition plays in consumer protection and in generating consumer benefits. This institutional change process phased in retail competition incrementally, which delayed customer benefits but also provided an environment in which customers and new entrants could have more confidence.

Policymakers in Texas recognized that retail competition was partly about price and using market processes to align costs and benefits more efficiently. They also realized, however, that it was about choice, product differentiation, and consumer empowerment that are impossible under retail regulation.

9

Market Monitoring, ERCOT Style

Andrew N. Kleit

The goal of restructuring is to create competitive markets for electricity, leading to the creation of wealth in society. In the Electric Reliability Council of Texas (ERCOT), as well as in other regional electricity wholesale market and system operators, a market monitor has been created to help ensure (or perhaps bring about) more competitive wholesale electricity markets. The creation of market monitors, however, raises a number of questions. In particular, the types of market behavior that should be controlled, and what mechanisms should be used to control behavior deemed inappropriate, are not clear. This chapter will review the establishment and functions of the ERCOT independent market monitor.

The Distrust behind Market Monitoring

In a classic article, the economist Harold Demsetz discussed the "trust behind antitrust."[1] In brief, the antitrust paradigm assumes that competitive markets are generally robust. That is, in general, open competition maximizes society's wealth. In such circumstances, antitrust's relatively small role is simply a complementary tool.

There is no such assumption behind wholesale electricity markets. Indeed, restructured wholesale electricity markets are quite possibly the only

---

I would like to thank Parviz Adib, Brad Jones, Clarence Johnson, Dan Jones, Lynne Kiesling, James Reitzes, and Eric Schubert for helpful comments, and Darren Bush for graciously sharing his ideas with me.

markets in the economy that have explicit market monitors. This is because, simply put, restructured electricity markets cannot be trusted to create competitive outcomes, at least not by themselves. As events and research have shown, electricity market competition is not robust.[2] There are several reasons for this.

First, unlike demand in other markets, demand for electricity is highly price-inelastic. This lack of responsiveness stems from the historic nature of analog metering of electricity at the end-user level.[3] In general, meters have simply recorded the total flow of electricity past a point during a certain time period. The result is that retail consumers pay for the average price of the electricity they consume, and most do not observe the wholesale price of electricity in their power bills.[4] Thus, in contrast to markets in other products, a firm or set of firms that is able to raise the wholesale price of electricity will suffer very little loss of sales due to the higher price. In their case, because exercising market power is more profitable, firms have larger incentives to create and exercise such power.

Second, the provision of electricity in any particular control area generally takes two different forms. The first relies on baseload facilities that have high fixed costs and low marginal costs, such as nuclear or coal plants. These plants cost a great deal to construct, but once they are built they run at low cost. The second form uses (to a large extent, in ERCOT) natural gas plants with low fixed costs and high marginal costs. These units run only when the price of electricity is relatively high. High wholesale prices generally occur during peak strains on the system resulting, for example, from air conditioning load during high summer temperatures. The result of all this is a supply curve for electricity that looks like a hockey stick. Figure 9-1 represents the supply curve for ERCOT at an arbitrary point in time.[5] As is clear from the figure, a rapid upward movement occurs when the price is about $150/MWh.

Third, the structure of generation plant ownership may create further incentives for the exercise of market power. As the discussion of "hockey-stick bidding" below indicates, if a firm has a combination of large baseload and small peaking plants, it may, in certain time periods, be able to raise the price it receives on its baseload plants by slightly reducing the output from its peaking plants.

Here is an example of the type of competitive problem that can occur in restructured electricity markets: Assume a firm has 1,000 MW of baseload

FIGURE 9-1
ERCOT BID STACK ACROSS 5 ERCOT ZONES,
JANUARY 31, 2007 (AM CST)

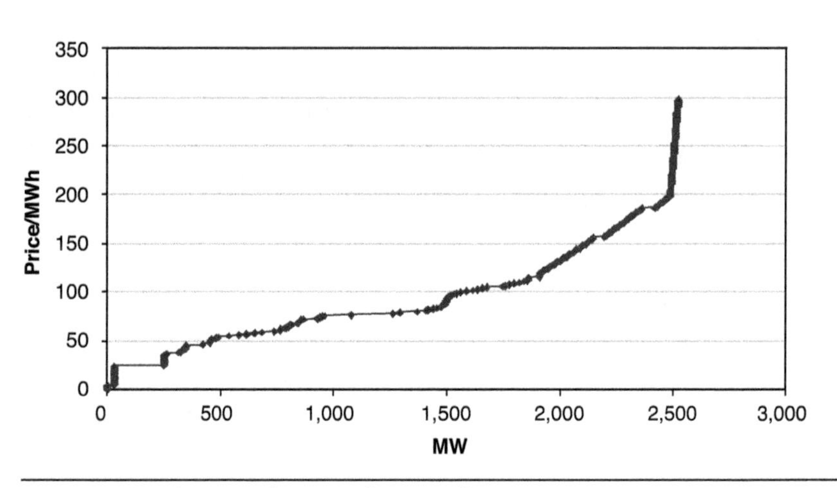

SOURCES: ERCOT data received courtesy of Steven Puller.

(say, coal) capacity that can produce power at a marginal cost of $50/MWh. Assume also that the firm has a "peaking" plant that can produce, if desired, 20 MW of power at a marginal cost of $200/MWh. There are many baseload plants operating at a marginal cost of around $50/MWh. So when the market price of power is, say, between $50 and $100 per MWh, our firm cannot affect the market price. It runs its baseload power plant, covering at least its marginal costs of operation, and does not run its peaking plant.

Assume, however, that on a particular day and time for an hour, the market price of power is $200/MWh. There are only a few peaking plants with marginal costs in that range. So if our firm decides not to run its peaking plant, it might be able to raise the market price to, say, $205/MWh. The firm will lose $100 on its peaking plant [(205–200)*20] but gain $5,000 (5*1,000) on its baseload plant.[6]

Note that this story does not depend upon a firm's share in a particular market, but rather upon more subtle factors, in particular the "hockey stick" shape of the market supply curve. Put another way, market power depends on whether a particular supplier is "pivotal" at a particular point and place in time, not on its overall market share. Thus, traditional market share indices used in antitrust, most notably the Herfindahl-Hirschmann Index, are not very useful in the wholesale electricity context[7]—a fact that, in turn, implies that regulation of market power in the electricity industry needs to be much more oriented toward specific situations than in the general antitrust area.

Another contrast between electricity and other markets, perhaps relating to the factors above, is the treatment of the existence of market power. In general antitrust, the mere exercise of market power is not considered an offense.[8] The basic rationale for this position is that firms gain market power because they are "good," or more efficient, at what they do.[9] Thus, those that have gained their market power by being more efficient should not be penalized with enforcement actions for their quest for efficiencies.

This assumption of efficiency clearly does not hold with respect to electricity markets for two reasons. First, as discussed above, the wholesale market in electricity is unlike most markets in that a firm does not have to be particularly large to exercise market power. Second, restructured electricity markets are the outgrowth of the regulatory process over several decades. Although the manner in which firms gained their existing market power is shrouded in the mists of that regulatory past, there is no reason to believe that they gained their positions through any type of economic efficiency. These factors imply that, contrary to the rule in antitrust courts, generators in wholesale electricity markets should not necessarily be permitted to exercise the market power that they have. Rather, some type of regulatory intervention may take place should they attempt, unilaterally, to influence the wholesale price of electricity.

It should be noted that market monitoring has taken on special importance in ERCOT due to a 2005 court ruling. In a case pitting retailer Texas Commercial Energy (TCE) against TXU, the U.S. Court of Appeals for the Fifth Circuit ruled that the "filed-rate doctrine" applied to ERCOT, despite the inconsistencies of such a position.[10] Application of the filed-rate doctrine means that the price of electricity in ERCOT cannot be challenged on antitrust grounds; this removes a potentially important pillar of the competitive market structure.

The Structure of Market Monitoring in ERCOT

Once a regulator and/or system operator has decided to establish a market monitor, several issues remain. First, the independence of the market monitor must be considered. Second, the authority of the market monitor must be established. This next section reviews the independence issue and discusses the evolution of the ERCOT independent market monitor (IMM). It then assesses the authority (or, perhaps, lack thereof) of the IMM and relates it to the observed problem with market power in ERCOT.

The Bowring Problem. In theory, a market monitor should be an independent analyst who reports the exercise of market power to a regional transmission organization (RTO). But the ability of the market monitor to be effective will be a function of who hires the market monitor, who pays the market monitor, and who controls what the market monitor can say to the public.

The five RTOs in the United States other than ERCOT have created their own office of market monitor, hired and funded by them. Most have hired outside consultants, although an outsider is disadvantaged by the difficulty of obtaining the wealth of real-time information it needs from the RTO. PJM, the RTO for the Mid-Atlantic region, however, hired its own internal market monitor, which became the subject of controversy in the spring of 2007. Dr. Joseph Bowring, the general manager of the PJM market monitoring unit, asserted in testimony before the Federal Energy Regulatory Commission (FERC) that PJM management was restricting the ability of the unit to submit comments in public forums about the competitiveness (or lack thereof) of various PJM markets.[11]

While the details are not important to this chapter, the basic dispute between Bowring and PJM appears to have been twofold. First, PJM management was concerned that the organization speak with "one voice," while Bowring felt he had an obligation to FERC to report his own independent judgment with respect to the state of PJM markets. Second, Bowring believed that PJM markets were less competitive than they might otherwise be, a view with which PJM management disagreed.

It should be noted that this type of disagreement is natural to the process of market monitoring. As the discussion below indicates, well-intentioned parties can have many disagreements about what constitutes a "competitive" market. Furthermore, in the case of PJM, many other parties had a stake in the state of the relevant market. Given this, Dr. Bowring's position that parties outside PJM management had a legitimate reason to have access to his analysis seems to have been a solid one.

But a second problem is perhaps more difficult. In theory, PJM, as an RTO, should also have been interested in competitive markets. Unfortunately, the governing mechanisms of an RTO are not well understood. PJM was, at least technically, governed by a board of directors comprising various stakeholders, including representatives of both producers and consumers. It is entirely conceivable, however, that the producers had a greater share of control than the consumers in the everyday workings at PJM.[12] The PJM staff was, in large part, made up of personnel who had previously worked for producing companies. Thus, PJM management may not have been as opposed to the exercise of market power as some might have wanted them to be.

The Bowring dispute made clear the tensions involved in being a market monitor. As the PJM special report put it,

> Answering two masters, the Market Monitor had responsibilities and obligations to FERC under its monitoring function pursuant to the PJM Tariff, while simultaneously, serving as an employee of PJM, having obligations imposed by PJM Management. These obligations to PJM Management did not always mesh seamlessly with the monitoring duties imposed by the FERC and the [managing board of PJM].[13]

Clearly, the PJM case shows us, a market monitor can no longer work inside an RTO and retain its full independence. Another solution is desirable.[14]

The Evolution of the ERCOT Independent Market Monitor. Unlike most other RTOs, ERCOT has the distinction of being wholly contained within one legal jurisdiction.[15] Thus, it was feasible for the Public Utilities Commission of Texas (PUCT) to create its own market monitor directly without the complications involved in multistate RTOs.

Originally, PUCT staff performed the market monitoring function, following the 1999 passage of Texas Senate Bill 7, the basic Texas restructuring law. One reason for a staff market monitor was major constituencies' skeptical view of the market monitor that was then inside the New England ISO and engaging in bid mitigation "by hand" in an apparently arbitrary manner.[16]

Funding for the staff market monitor came originally from the PUCT and the state legislature. The legislature, however, was unwilling to continue using public funds for this purpose. Senate Bill 408, passed in 2005, codified the legal status of the market monitor, with the function moving from the PUCT to an independent market monitor. As directed by SB 408, the IMM was funded through a fee imposed by ERCOT, although it was selected not by ERCOT, but by the PUCT. After receiving a series of bids, the PUCT hired Potomac Economics to be the independent market monitor. Potomac Economics was also the IMM for the New England ISO, the New York ISO, and the Midwest ISO. The IMM began full-time operation in ERCOT in October 2006.

To some extent, the ERCOT IMM is subject to pressure from ERCOT. Its office is located in ERCOT's offices in Austin, and it requires real-time data from ERCOT, which the organization is required to supply. The legal mandate to supply such data seems unclear and is, perhaps, unenforceable. A meeting beween the author and the director of the IMM unit in November 2007 did not reveal any tension between the IMM and ERCOT.[17] The true test will come, however, when the two have a serious disagreement.

Market Monitoring and Bid Mitigation. In the other RTOs in the United States, the RTO, working with the market monitor, engages in automatic bid mitigation, referred to as "automatic mitigation procedures" (or AMP). AMP works something like this: Each plant's marginal cost of producing electricity is calculated using the relevant formula. For example, for natural gas units, the "crude" marginal cost can be calculated by multiplying the heat rate by the price of natural gas. If the heat rate on a unit is, say, 8 MWh, and the price of natural gas is $7/mmBTU, then the "crude" marginal cost of that unit is 8*7=$56/MWh.[18]

After the crude marginal cost is calculated, a "reasonable" profit margin is added to create an acceptable bid price. For example, assume that a generation unit has an estimated "crude" marginal cost of $10/MWh. Assume also that the "reasonable" profit acceptable to the ISO is 15 percent. In these circumstances, the ISO will accept a bid on that unit of up to $10*1.15=$11.50. Any bids above that level are "mitigated" down to $11.50.

Submitting bids above marginal cost to raise the market price is referred to as "economic withholding," and it is prevented by AMP. Firms may, however, exercise market power by engaging in "physical withholding," by not submitting bids and refusing to make their generation facilities available at any price. Such behavior is not addressed by AMP, and it is unclear how it would be addressed by a market monitor.

The choice of the proper role for a market monitor represents the tension between two types of mistakes. The first mistake is to be "too lenient" by allowing the exercise of market power. The second (which may be more subtle) is to be "too strict" by forcing prices to reach "competitive" levels, thereby discouraging "dynamic efficiency" by deterring new entry into the market. Avoiding the first type of mistake requires setting mitigation levels close to marginal cost, while avoiding the second implies leaving a large "safety zone," or not engaging at all in AMP. In the end, a choice between these two types of mistakes relies largely on the judgment of the policymakers.

Complicating matters further, the true economic marginal cost may not be that easy to measure. Determining the appropriate marginal cost may be more involved than performing the calculation of multiplying the heat rate by the price of natural gas. Brennan points out that many electricity generating plants have to be started and stopped in real time.[19] This process has real costs and makes calculating the "marginal cost" of production for these plants somewhat problematic. In addition, plants are periodically scheduled for maintenance. During (or just prior) to these times, it may well make sense to delay maintenance, if the market price is high enough. For these plants and times, true opportunity cost includes the cost of deferred maintenance, the calculation of which again is not straightforward.

Moreover, AMP programs do not seem well-formulated toward the specific exercise of market power. As the discussion above indicates, opportunities to exercise market power in wholesale electricity markets are not simply a function of firm size. They depend critically on what assets are in a firm's generation portfolio and what the demand conditions are at a given point in time. AMP programs that are not fine-tuned to these distinctions risk engaging in mitigation when it is not appropriate, reducing the dynamic efficiency of the relevant market.

Perhaps alone among ISOs, ERCOT has not authorized (or had the PUCT direct it to authorize) its market monitor to engage in automatic bid mitigation. The rationale for this decision appears to be that because ERCOT does not have a capacity market, it does not want to make the mistake of discouraging new investment.[20]

The PUCT did, in fact, implement one AMP program. In 2003, it implemented a bid-mitigation program, apparently designed to deal with the "hockey stick bidding" that occurred during an ice storm that February.[21] The impacts of the ice storm and the resulting AMP program are discussed below.

Evidence of Market Behavior in ERCOT. Two studies have reviewed the behavior of firms in the ERCOT balancing (day-ahead) market. Hortaçsu and Puller, reviewing events in 2001–2, and Sioshansi and Oren, reviewing events in 2002–3, came to very similar conclusions.[22] Both revealed that large firms exercise unilateral market power when they have the opportunity. While there is no evidence of any type of explicit or tacit collusion, these firms appear to understand the circumstances under which they can profit by submitting bids above marginal costs.

According to the studies, small firms, too, submit bids that are unequal to their marginal costs—a surprising finding, in that these firms are not in a position to exercise market power. More precisely, the small firms actually submit bids that deviate *too far* from marginal cost. As they are both buyers and sellers of power that conduct most of their decisions in other markets, such firms lack incentive to worry about what their bids should be. Both studies concluded that the PUCT should do more to encourage them to pay greater attention to the balancing market.[23]

The 2003 Ice Storm and the Resulting Bid Mitigation Program

The difficulties of market monitoring were made apparent in February 2003, when an ice storm and cold temperatures covered most of the state of Texas. The storm cut off natural gas supplies, reducing the supply of electricity generation. Adding to the problem was ERCOT's substantial underestimation of the actual demand, which left many generation units in the middle of maintenance when they were needed.

As a result, the price of balancing power in Texas soared during this period. Though the internal market monitor report is not specific, balancing market prices appear to have averaged approximately $350/MWh during February 24–26. This compares with prices averaging in the $25–$40/MWh range for the rest of 2003. At times during these three days, the price reached $999/MWh. The May 2003 market monitor report indicated that one (unnamed) supplier engaged in "hockey stick bidding" (although the January 2004 PUCT market monitor report said that it was not TXU, the largest supplier and therefore, perhaps, the most likely suspect).[24]

As a result of the high prices during the February 2003 ice storm, the PUCT adopted a limited AMP program for ERCOT. In brief, the commission offered a three-step mitigation program. The first step was to discover whether the "bid stack" was completely used. If so, one could conclude that the competitive market no longer "worked," and that another approach would be appropriate. Second, all plants with marginal costs at or below the 95 percent level in the bid stack would receive the ninety-fifth-percentile marginal cost times 1.5.[25] Third, all plants with marginal costs above that level would be paid at the price they bid. In addition, any bids above $300/MWh would be publicly announced, creating what some referred to as a "wall of shame."[26]

The ice storm analysis produced by the internal market monitor also foreshadowed the PUCT staff's filing of an administrative complaint against TXU. The report found that TXU was a pivotal supplier 17 percent of the time in 2003.[27] As the internal market monitor staff wrote, "The results of this study show that TXU's market position is so pivotal that just about anything the company does with respect to the balancing market will affect balancing energy prices, regardless of the reasons behind its decision."[28]

The Texas AMP program was eliminated by the PUCT in October 2006, largely due to events that occurred on April 17, 2006. An unusually warm day resulted in increased demand for power across the ERCOT region. AMP mitigation was activated, and prices were lowered to around $200/MWh. Unlike what had happened during the February 2003 ice storm, however, rolling blackouts occurred across the state. While there was no direct evidence that the price mitigation had reduced supply, the PUCT apparently believed that supply reductions were possible. Moreover, the PUCT apparently believed that the AMP program was restricting firms gaining "inframarginal rents" (profits that occur when firms offer their products at marginal cost when the market price is above the marginal cost), rather than benefiting from market power. To the extent that was correct, the AMP program was reducing welfare-enhancing entry into the market.[29]

The TXU Matter

Against the urgings of PUCT staff, TXU, the largest utility in ERCOT, began what it called its "Rational Bidding Strategy" (RBS), starting in late 2004. As TXU explained, a firm has both variable and fixed costs. Variable costs can be recovered through marginal-cost pricing. But marginal-cost pricing is not sufficient to cover fixed costs, at least if each plant is allowed to be reimbursed solely on its marginal costs. Therefore, TXU announced, it would be including some fixed costs as well as variable costs in its bid offers for various production facilities in the "real-time" or "balancing" market in ERCOT.

The Trouble with the RBS. For neoclassical economists, the RBS strikes a distinctly negative cord. According to the theory of competitive markets found in any introductory economics textbook, firms are willing to offer their products at the marginal cost of producing them. The market price becomes the marginal cost of the last unit produced—that is, the unit needed to "clear" the market and set the number of units sold equal to the number of units purchased. As discussed above, the difference between firms' marginal costs and the market price is referred to as the "inframarginal rent," and it covers the fixed costs of production. Electricity firms gain their inframarginal rents during times when demand causes prices to rise above their marginal costs. There is no need in such markets to make offers for production above their marginal costs.[30]

With solely this information, it is not clear that TXU would be committing a market power infraction by abandoning marginal-cost pricing. If it did so through a basic misunderstanding, it would lose a great deal of money by overpricing its generation units during competitive periods. While such a strategy would also be harmful to consumers, that might not be sufficient for a legal case. There is no legal rule that states that TXU or any other firm has to behave in full concordance with economic theory.

What fatally undercut this argument, however, is evidence uncovered by the independent market monitor that the firm did not engage in the RBS in all periods. According to a March 2007 IMM report, TXU stated to the PUCT that

> in some periods, market conditions appear to make it reasonably unlikely that the units will be called upon with bids representing full costs. In these periods, it may not be economically efficient to bid at a price reflecting the full cost of owning, operating, and maintaining the unit, since no revenue at all will be received if the bid is not taken and, to some extent, any recovery of cost above short run marginal cost is better than no recovery.[31]

In other words, TXU engaged in the RBS only during periods when doing so would raise the market price to TXU's benefit. Thus, the RBS was designed for and achieved the exercise of market power.

Using the March 2007 IMM report as a basis, the PUCT staff filed an administrative complaint before the commission in that month, asking for $70 million in refunds from TXU and $140 million in administrative fines.[32] An amended complaint, filed in September of that year, reduced the plea for fines and penalties to a total of $171 million.[33]

The TXU Response. In its response to the PUCT complaint, TXU made a number of important points. Many of its comments dealt with the same issues examined by Bush and Mayne in a 2007 study of what makes antitrust enforcement difficult in electricity markets: The product market is difficult to define; concentration measured by the Herfindahl Hirschman Index does not reach levels considered "threatening" in other markets; and, where traditional antitrust cases must deal with a "significant" duration of anticompetitive behavior, the duration in electricity markets may be alleged in terms of hours or even minutes.[34]

Perhaps the most important rebuttal, however, was TXU's critique of the theory the PUCT staff was applying to the case. The staff was not claiming that TXU colluded with any party, or precluded any party from entering the market. Rather, contrary to established antitrust doctrine, it was bringing a complaint against TXU for the simple unilateral exercise of market power.[35]

Whether or not such simple exercise of market power is against Texas law is highly unclear. According to PUCT Substantive Rule § 25.504(b)(3), market power abuse is defined as

> [p]ractices by persons possessing market power that are unreasonably discriminatory or tend to unreasonably restrict, impair, or reduce the level of competition, including practices that tie unregulated products or services to regulated products or services or unreasonably discriminate in the provision of regulated services. Market power abuses include predatory pricing, withholding of production, precluding entry, and collusion.[36]

Section D goes on to define "withholding of production":

> Prices offered by a generation entity with market power may be a factor in determining whether the entity has withheld production. A generation entity with market power that prices its services substantially above its marginal cost may be found to be withholding production; offering prices that are not substantially above marginal cost does not constitute withholding of production.[37]

Thus, the PUCT's rule appeared to make clear that the PUCT's staff was truly alleging "abuse of market power" against TXU.

The PUCT's rule was, however, subject to judicial review, and the definition of abuse of power, as TXU pointed out, was an important part of that review. The rule was upheld by the Austin Appeals Court in *TXU v. PUCT*.[38] Unfortunately, the court's actual statement on the issue was unclear:

> In light of the administrative record and the explicit statements by the Commission that the [market power abuse] Rule does not require marginal-cost pricing, we find no merit in the theory that a market participant will be coerced into setting prices at an unprofitable rate out of an unsubstantiated fear that the Commission will initiate an investigation.[39]

Regrettably, there is no record of what the commission said on this matter, or whether the appeals court was simply rebutting the flawed economics behind TXU's "rational bidding strategy." At this point, however, it seems an open question whether or not the simple exercise of power constituted "market power abuse" in ERCOT.

Along these lines, TXU pointed out that the PUCT staff assessed its plea for damage payments based on the difference between the market price and the marginal cost in the market.[40] Thus, the staff was implicitly claiming that there was no margin above marginal cost that TXU could legally charge.

TXU's Voluntary Mitigation Plan. In July 2007, TXU offered the PUCT a voluntary plan to mitigate its market behavior (voluntary mitigation plan, or VMP), and this plan was accepted by the commission in August 2007 in the form of an uncontested proceeding.[41] While the details of the proposed VMP were somewhat complex, the rules for the balancing market were relatively clear.

For TXU's gas-fired units (the bulk of TXU's generation), the first 75 percent of its bid offers would be at a price not higher than the product of a 12.5 heat rate (mmBtu/MWh) and the price of natural gas. The next 22 percent would be at a price not higher than the product of a 15 heat rate (mmBtu/MWh) and the price of natural gas. The final 3 percent of TXU's generation would be offered at a price not higher than the low system offer cap (generally $500). The plan was originally designed to be in place for one year.

The VMP becomes more understandable if its provisions are presented in tabular form. Assume that the price of natural gas is $8/mmBTU. In this case, the rate caps offered by TXU on its generation would be as shown in table 9-1.

It is difficult to know exactly what problem the VMP would be solving.[42] While market prices of $100/MWh are not unheard of, they are not typical, either. Puller reports that prices above $100/MWh occurred about 4.7 percent of the time in 2006.[43] Further, only a small amount of capacity is needed to engage in "hockey-stick" bidding, as discussed above. It may well be that 3 percent of TXU's capacity could be enough to execute such a strategy successfully. Of course, the VMP would have allowed TXU to price above marginal cost, contrary to the theory apparently underlying the PUCT staff's complaint against the firm.[44]

TABLE 9-1
TXU's VOLUNTARY MITIGATION PLAN, ASSUMING A NATURAL GAS PRICE OF $8/MMBTU

Fraction of TXU's generation capacity	Allowed heat rate	Allowed bid price $/MWh
First 75 percent	12.5	$100
Next 22 percent	15	$120
Remaining 3 percent	–	$500

Perhaps regrettably, the procedure by which the PUCT adopted TXU's VMP was ruled inappropriate in October 2007.[45] The court imposed an injunction on the VMP, ruling that the commission needed to review it as a contested, rather than an uncontested, matter. This would involve longer proceedings and the possibility of extensive discovery on the moving party (here, TXU). Thus, the VMP has not gone into effect, and there is no sign that TXU wishes to pursue the matter.

Conclusion

While problems may still lie ahead, the PUCT has created an independent market monitor that has perhaps the best potential to remain independent and effective. The IMM is an integral part of the ERCOT structure, but it reports directly to the PUCT. While events could put stress upon this arrangement, it appears far better than the arrangements at RTOs such as PJM that reach across several states.

If the PUCT is serious about market mitigation, however, it needs to adopt a broader understanding of the potential for the exercise of market power in ERCOT. In particular, it needs to understand that, given the right circumstances, almost all generating firms in ERCOT may be able to exercise market power. It also needs to apply clear economic rationale to any market-monitoring policies. In particular, it is difficult to see what problem the abortive 2007 consent decree with TXU would have been addressing.

Devising a consistent approach for the control of market power is a far more challenging issue for the PUCT. In my view, the commission and/or the state of Texas have three different coherent strategies open to them.

First, consistent with antitrust principles, the PUCT could permit the unilateral exercise of market power. This strategy would require dismissal of the TXU case and, quite possibly, would result in an increase in ERCOT electricity prices, at least in the short run. The hope would be that new entry into electricity generation would dilute the ability of TXU and other firms to exercise market power. Such a policy would have two important potential flaws. Unlike in most industries in the antitrust context, it is not clear that TXU and other firms have "earned" their positions of market power. Rather, they appear to have acquired them as a result of the regulatory process. Further, given the sensitive nature of electricity markets, new entry may not significantly alleviate the market power problem in the relatively near future.

Second, ERCOT could adopt active market monitoring and bid mitigation, as have the other RTOs. This approach would have the advantage of controlling the exercise of market power directly, unlike the cumbersome administrative process the PUCT is currently directing at TXU. It would, however, also pose problems. For one thing, it might induce ERCOT prices that are too low to permit wealth-creating entry into generation markets. For another, it still would not directly address the issue of physical withholding, where firms simply choose not to offer their plants for generation at any price.

Finally, the PUCT could appeal to the Texas legislature for authority to engage in divestiture of plants from firms that are in a position to exercise market power. Such a proposal would be sure to create opposition from potentially affected firms, and the divestiture would have to be carefully directed to eliminate plant combinations that could create market power, a situation unique to electricity markets. Such a precise divestiture program has not previously been tried, and there is no guarantee that such a subtle regulatory process could be successful.

Each of these approaches has its weaknesses. But adopting any of them would address the logical inconsistency in the ERCOT market-monitoring program, an inconsistency that is likely to be a recurring problem in the regulation of the ERCOT market.

Notes

Preface

1. Peter Van Doren and Jerry Taylor, *Rethinking Electricity Restructuring*, Cato Institute Policy Analysis No. 530, November 30, 2004, available at http://www.cato.org/pub_display.php?pub_id=2609 (accessed October 7, 2008).

Introduction

1. See, for example, Rebecca Smith, "Deregulation Jolts Texas Electric Bills," *Wall Street Journal*, July 17, 2008, A1.

2. Centerpoint/Reliant was the incumbent utility in the Houston area that, under restructuring, became the regulated transmission and distribution service provider (TDSP), or wires company, for that area. Similarly, TXU became the TDSP for the Dallas–Fort Worth area and subsequently changed its name to Oncor.

Chapter 1: Why Does ERCOT Have Only One Regulator?

1. 273 U.S. 83 (1927). The Supreme Court's so-called dormant commerce clause jurisprudence interprets the Constitution's grant of power to regulate interstate commerce as implying a restriction on the states' power to do so. The Supreme Court has adhered to this view ever since *Gibbons v. Ogden,* 22 U.S. 1 (1824).

2. The FPC had some authority prior to 1935 over utilities, but that power was greatly expanded under the 1935 amendments. See Federal Water Power Act of 1920, ch. 285, 41 Stat. 1063, amended by the Federal Power Act of 1935, ch. 687, 49 Stat. 863 (codified as amended in scattered sections of 16 U.S.C. (1976)).

3. Federal Power Act, 16 U.S.C.A. § 824(b)(1).

4. 16 U.S.C.A. § 824(c).

5. 16 U.S.C.A. § 824(b)(2).

6. FERC has confirmed its conclusion that "intrastate transmission," as defined in the FPA, applies only to states like "Alaska, Hawaii, and most areas of Texas" that "have no interconnection that would permit the physical transmission of power outside of a State." See Brief for the Federal Energy Regulation Commission in Opposition to a

Writ of Certiorari at 13 n. 9, *New York v. Fed. Energy Reg. Comm'n*, 531 U.S. 1189 (2001) (No. 00-568).

7. In *Connecticut Light and Power v. Federal Power Commission*, 324 U.S. 515 (1945), the Supreme Court held that these specific reservations constituted a "legal or governmental standard" that must be given effect, in addition to the technological transmission test, in assessing jurisdictional status. See also *Fed. Power Comm'n v. Southern Cal. Edison*, 376 U.S. 205, 215 (1964).

8. Cassandra Robertson, *Bringing the Camel Back into the Tent: State and Federal Power over Electricity Transmission*, 49 Clev. St. L. Rev. 71, 78 (2001). See also *Wickard v. Fillburn*, 317 U.S. 111, 127–28 (1942), holding that even wheat grown and consumed on a single farm substantially affects interstate commerce because that consumption of wheat decreases the demand for wheat in general. In *Federal Energy Regulatory Commission v. Mississippi*, 456 U.S. 742 (1982), a rare constitutional challenge was brought against the Public Utilities Regulatory Policies Act (PURPA; see below), but the Supreme Court reiterated the broad scope of the commerce clause in the energy context.

9. *Connecticut Light & Power Co.*, 324 U.S. at 529.

10. Richard C. Cudahy, "The Second Battle of The Alamo: The Midnight Connection," *Natural Resources and Environment* 56 (Summer 1995).

11. Ibid.

12. The antitrust action *West Texas Utilities Co. v. Texas Electric Service Co.*, 470 F. Supp. 835 (N.D. Tex.1979) paints a rich picture of ERCOT, describing a voluntary membership organization where the obligation not to transmit interstate was often not formalized but "understood."

13. Tex. Rev. Civ. Stat. Ann. art. 1446c (Vernon Supp. 1992).

14. The movement in Texas toward public utility regulation was no easy sledding. See Jack Hopper, *A Legislative History of the Texas Public Utility Regulatory Act of 1975*, 28 Baylor L. Rev. 777 (1976).

15. The Supreme Court, in *New York v. Federal Energy Regulatory Commission*, 535 U.S. 1, 7–8 (2002), offered its sense that "it is only in Hawaii and Alaska and on the 'Texas Interconnect'—which covers most of that State—that electricity is distributed entirely within a single State. In the rest of the country, any electricity that enters the grid immediately becomes a part of a vast pool of energy that is constantly moving in interstate commerce."

16. This expansiveness is well illustrated by *New York v. Federal Energy Regulatory Commission*, 535 U.S., at 1. The Court noted that there have been sweeping changes in the electric power industry since passage of the FPA. The Court then stated that, because of those changes it was "left with the statutory text [of the FPA] as the clearest guidance. That text unquestionably supports FERC's jurisdiction to order unbundling of wholesale transactions (which none of the parties before us questions), as well as to regulate the unbundled transmissions of electricity retailers." Id. at 23–24. Other illustrative cases include *Federal Power*

Commission v. Florida Power and Light Co., 404 U.S. 453 (1972), which found transmission in interstate commerce when the power of an intrastate utility was "commingled" with power being sent out of state, and *Jersey Central Power & Light Co. v. Federal Power Commission*, 319 U.S. 61, 71–72 (1943), which found transmission in interstate commerce when an intrastate utility transmitted power to an unaffiliated entity in the same state, which then transmitted that quantum of energy across state lines.

17. See generally Cudahy, "Second Battle of the Alamo."

18. See *West Texas Util. Co. v. Texas Elec. Serv. Co.*, 470 F. Supp. 835 (N.D. Tex. 1979); *West Texas Util. Co. v. Texas Elec. Serv. Co.*, 470 F. Supp. 798 (N.D. Tex. 1979) (denial of motion for new trial).

19. The Federal Power Act provides that where jurisdiction exists and is in the public interest, the FPC may order interconnection, pursuant to 16 U.S.C. § 824a(b) and 16 U.S.C. § 824a(c). For a good description of the events surrounding the "midnight connection," see Cudahy, "Second Battle of the Alamo," 6.

20. *Central Power and Light v. Fed. Power Comm'n*, 56 F.P.C. 432 (1976). The FPC found a reasonable probability that some systemic disruption would arise due to the now bifurcated nature of the energy grid, and thus that an emergency situation was present under 16 U.S.C. § 824a(d). This determination allowed the FPC to order temporary connection during emergencies for entities *not* otherwise subject to FPC jurisdiction. So finding, the FPC ordered the disconnected Texas utilities to restore temporarily certain important physical connections, emphasizing that such forced interconnections were not to impart jurisdictional status. Due to lack of clarity in its decisional underpinnings, the FPC judgment was remanded for reconsideration by the Circuit Court of Appeals for the District of Columbia in *Central Power and Light Co. v. Federal Energy Regulatory Commission*, 575 F.2d 937 (D.C. Cir. 1978), *cert. denied* sub nom., 439 U.S. 981 (1978).

21. See *Application of Houston Lighting and Power Co. for Reconnection of the Texas Interconnect System*, P.U.C.T. Docket No. 14 (July 11, 1977).

22. Pub. L. No. 95-617 (codified at 16 U.S.C. §§ 796-825r (1976 & Supp. V 1981)).

23. For an explanation of wheeling, see chapter 6, note 19.

24. Under section 205(a) of PURPA, FERC was granted the authority to exempt "electric utilities" from any state law, rule, or regulation that "prohibits or prevents the voluntary coordination of *electric utilities*," should the "Commission determine that such voluntary coordination is designed to obtain economical utilization of facilities and resources in any area." 16 U.S.C. § 824a-1(a) (emphasis added). At the same time, sections 202, 203, and 204 of PURPA authorized FERC to order interconnection and wheeling where statutory conditions are met: "Upon application of any *electric utility* . . . the Commission may issue an order requiring . . . the physical connection of . . . the transmission facilities of any *electric utility*, with the facilities of such applicant." See 16 U.S.C. § 824i, 824j, and 824k (emphasis added).

25. Offer of Settlement in *Central Power and Light*, Docket No. EL79-8. See *Central Power and Light Co.*, 17 F.E.R.C. ¶ 61,078 (1981).

26. *Order Requiring Interconnection and Wheeling, and Approving Settlement*, 17 F.E.R.C. ¶ 61,078 (October 28, 1981).

27. Ibid., 6.

28. Pub. L. No. 102-486, 106 Stat. 2776 (1992).

29. See 16 U.S.C. § 824i–j.

30. 16 U.S.C. § 824j. This provision replaced the phrase "electric utility" employed by PURPA with "transmitting utility" to broaden coverage of the intended class. This provision was again amended by the EPAct of 2005; see below.

31. Under § 824k(k), "Any order under section 211 requiring provision of transmission services in whole or in part within ERCOT shall provide that any ERCOT utility which is not a public utility and the transmission facilities of which are actually used for such transmission service is entitled to receive compensation based, insofar as practicable and consistent with subsection (a), on the transmission ratemaking methodology used by the Public Utility Commission of Texas." 16 U.S.C. § 824k(k).

32. Order No. 888, Promoting Wholesale Competition Through Open Access Non-discriminatory Transmission Services by Public Utilities; Recovery of Stranded Costs by Public Utilities and Transmitting Utilities, F.E.R.C. Stats. & Regs. ¶ 31,036, 61 Fed. Reg. 21,540 (1996) (hereafter Order 888), order on reh'g, Order No. 888-A, F.E.R.C. Stats. & Regs. ¶ 31,048 (1997) (codified at 18 C.F.R. pts. 35, 385, order on reh'g, Order No. 888-B, 81 F.E.R.C. ¶ 61,248, 62 Fed. Reg. 64,688 (1997), order on reh'g, Order No. 888-C, 82 F.E.R.C. ¶ 61,046 (1998), aff'd in relevant part sub nom., *Transmission Access Policy Study Group v. FERC*, 225 F.3d 667 (D.C. Cir. 2000), aff'd sub nom., *New York v. FERC*, 535 U.S. 1 (2002).

33. As FERC stated in Order No. 888, footnote 516, "The Commission also has jurisdiction to order wholesale transmission services in either interstate or intrastate commerce by transmitting utilities that are not also public utilities." In making the statement, FERC specifically cited an application of jurisdiction to ERCOT. See *Tex-La Electric Cooperative of Texas Inc.*, 67 F.E.R.C. ¶ 61,019 (1994), where, under FPA sections 211 and 212, 16 U.S.C. § 824j–k, FERC ordered wholesale transmission services to a rural electric cooperative operating within ERCOT. See also *Kiowa Power Partners LLC*, 99 F.E.R.C. ¶ 61,251 (2002), sustaining federal jurisdiction over ERCOT entities. At that juncture, there was no doubt that "Sections 211 and 212 of the FPA clearly give this Commission jurisdiction to order transmission services within ERCOT, subject to the special rate provision for ERCOT utilities in 212(k)." *City of College Station*, 76 F.E.R.C. ¶ 61,138 (1996).

34. 16 Tex. Admin. Code §§ 25.191–98 (2005) (original version at 16 Tex. Admin. Code §§ 23.67–70).

35. See Tex. Util. Code Ann. §§ 31.001–41.104 (Vernon 1999 & Supp. 2000); Robert J. Michaels, *Electricity in Texas,* Report of the Texas Public Policy Foundation

(Feb. 2007); see also Public Utility Commission of Texas, Sunset Staff Report, appendix C, available at www.sunset.state.tx.us/79threports/puc/puc.pdf.

36. More recently, the Energy Policy Act of 2005 (EPAct of 2005) changed the federal jurisdictional landscape yet again by amending FPA section 211. The term "transmitting utility," previously defined in that section to encompass "any transmitting entity," was redefined to an entity that "owns, operates, or controls facilities used for the transmission of electric energy (a) in interstate commerce and (b) for the sale of electric energy at wholesale." This does not change the basic accommodation between FERC and ERCOT. 16 U.S.C. § 796, as amended by EPAct of 2005, Pub L. No. 109-58, § 1291, 119 Stat. 594, 984.

37. *Brazos Electric Power Cooperative Inc.*, 118 F.E.R.C. ¶ 61,199 (2007). Brazos's "Offer of Settlement" was explicitly designed to obtain a FERC order "pursuant to Sections 210, 211 and 212 of the FPA directing the entities to provide the needed transmission and interconnection while not subjecting entities in the ERCOT grid, as well as ERCOT itself, to the jurisdiction of the FPA for any purposes other than the carrying out of the provisions of Sections 210, 211 and 212 of the FPA." *Brazos Electric Power Cooperative, Offer of Settlement,* P. 8. The unique question involves FERC's amended jurisdiction under FPA section 211, 16 U.S.C. §824j. Under section 210, unchanged by the EPAct of 2005, interconnection can still be sought by "any electric utility" with "any cogeneration facility" or "any electric utility." Under section 211, however, the new definition of "transmitting utilities" had to be satisfied. Although the entities did not buy or sell electricity, they owned and operated facilities used for its transmission in interstate commerce, and for its sale at wholesale, as a result of the commission directives establishing the two preexisting interconnections to which they were linked. Thus meeting the criteria of being a "transmitting entity," each could be subjected to jurisdictionally neutral FERC orders to wheel and interconnect with each and the SPP, and the third interconnection could move forward while ERCOT as a whole maintained its fundamental status.

38. Richard Cudahy described the deregulatory impulse this way: "The strong push, primarily of large industrial customers (and these were the real force behind deregulation), was for retail competition (competition for end users). To large industrial users, cheaper power was worth fighting for." Richard Cudahy, *Whither Deregulation: A Look at the Portents*, 58 N.Y.U. Ann. Surv. Am. L. 159 (2001).

39. The California restructuring plan was embodied in legislation (Assembly Bill 1890). In a series of orders issued during 1996 and 1997, FERC approved the restructuring proposals, which called for the three major public utilities in California—Pacific Gas and Electric Company (PG&E), Southern California Edison Company (SCE), and San Diego Gas & Electric Company (San Diego)—first, to transfer operational control of their respective transmission systems to the California independent system operator (CalISO), and second, to purchase all of the energy needed to serve their retail customers through wholesale spot markets. (These were day-ahead or day-of markets administered by the California Power Exchange Corporation, or "Cal

PX"). The three public utilities were precluded by California from entering into long-term contracts and were required to make all their purchases (and sales) through Cal PX's spot markets. Also, each utility's retail rates were frozen by California statute until it had earned enough revenue to recover certain costs associated with the transition to competitive markets. These are generally cited as major flaws in California's ill-fated attempt at restructuring. See, for example, Timothy J. Considine and Andrew N. Kleit, "Can Electricity Restructuring Survive? Lessons from California and Pennsylvania," in *Electric Choices: Deregulation and the Future of Electric Power*, ed. Andrew N. Kleit, 39–62 (London: Rowman and Littlefield, 2006).

40. Many states used some form of price freeze during the transition to markets, though the particulars varied considerably. In Texas, for example, incumbent utilities opening their service areas to competition were prohibited from lowering their prices during the transition period, so as to permit new entrants to get a foothold in the market. This so-called price-to-beat approach stood in contrast to California's state-imposed ceiling on retail rates during the transition period, prohibiting incumbent utilities from raising their rates. This ceiling exacerbated the electric rate crisis that befell Californians in the winter of 2000–1 by blocking normal demand responses to high (wholesale) prices.

41. This is the average unconstrained price as reported by the California Power Exchange. The Cal PX short-term and spot markets worked like a clearinghouse system, with a day-ahead market and a real-time market. For example, in the day-ahead market, sellers would submit bids indicating how much power they would be willing to sell into the system the next day and at what prices. Similarly, buyers would submit bids indicating how much power they would be willing to buy and at what prices. The Cal PX matched up these sell and buy bids to "clear" the market, while the Cal ISO and PX worked together to ensure that the transmission system could accomplish the delivery of power from sellers to buyers. Information from the day-ahead market was also used to arrange for the availability of power generation reserves, to ensure enough power would be available each day to serve sudden increases in demand. Early market operations proceeded relatively smoothly, with average wholesale energy prices at levels below those previously experienced in a cost-based regulatory regime—about $33/MWh for the first two years, compared with about $50/MWh before then. But the Cal ISO eventually experienced problems, leading to the imposition of a $750/MWh purchase-price cap (that is, the Cal ISO would reject offers to sell power to it at prices above this level). In May 2000, real-time prices in the Cal PX market reached the Cal ISO's $750 wholesale price cap (for more on this, see below) for the first time, and the Cal PX *average* price in its day-ahead market for the month topped $316/MWh. In June 2000, prices reached levels that exceeded by three or four times those seen in comparable demand conditions in prior years. Thus began what has been termed the "California energy crisis."

42. Federal Energy Regulatory Commission, *Final Report on Price Manipulation in Western Markets*, ES-1 (March 2003).

43. Public Citizen, Largest Fines, Penalties and Refunds Ordered by Federal and State Authorities against Corporations for Manipulation of the West Coast Energy Market and Natural Gas Price Index Manipulation, available at http://www.citizen.org/documents/camarketfines.pdf.

44. In actuality, the price caps were many, and they varied greatly as California scrambled to regain hold of its markets.

45. One measure of the failure of the price cap might be how often it was changed. The initial price cap was $250/MWh in the ISO real-time market, which was raised to $750/MWh on September 30, 1999, and reduced to $500/MWh on July 1, 2000. On August 7, 2000, price caps were reduced once more to $250/MWh.

46. See Debora Raggio Bolton, *Restructuring of the Electric Utility Industry,* 1274 PLI/Corp 303, 326 (2001).

47. See Darren Bush and Carrie Mayne, *In (Reluctant) Defense of Enron: Why Bad Regulation Is to Blame for California's Power Woes (or Why Antitrust Law Fails to Protect against Market Power When the Market Rules Encourage Its Use)*, 83 Or. L. Rev. 207 (2004).

48. 95 F.E.R.C. ¶ 61,418, 62,549.

49. The State of New York challenged FERC's claim of jurisdiction over retail wheeling, but the U.S. Supreme Court rejected New York's claim, reasoning that retail wheeling was permissible where retail transmission sufficiently affected interstate commerce. *New York v. Fed. Energy Regulatory Comm'n,* 535 U.S. 1 (2002).

50. *Keogh v. Chicago. & Nw. Ry. Co.*, 260 U.S. 156, 160–62 (1922) (holding that rates approved by the ICC cannot support individual damages under section 7 of the Sherman Act, even if collusive, citing Antitrust Act, ch. 647, § 7, 26 Stat. 209 [1890] [current version at 15 U.S.C. § 7 (2000)]).

51. That is, it lacks primary jurisdiction in any meaningful sense. Of course, in some cases, the conduct that might be alleged is *solely* the setting of the rate in question. In such a circumstance, the agency's determination of the rate as just and reasonable might very well end the issue. In most cases, however, the rate-setting is tangential to some other conduct (for example, the tying of a regulated service subject to a tariff to another service). In that case, the agency's approval of the rate is unlikely to be determinative. Moreover, the agency action might not even be informative to the court. See Darren Bush, *Mission Creep: Antitrust Exemptions and Immunities as Applied to (De)regulated Industries*, 3 Utah L. Rev. 613 (2006).

52. See Symposium, *Creating Competitive Wholesale Energy Markets*, 1 Envtl. & Energy L. & Pol'y J. 1 (2006).

53. 384 F.3d 756, 762 (9th Cir. 2004).

54. Ibid., 761 (quoting 16 U.S.C. §§ 824d(a), (c)) (citation omitted).

55. Ibid.

56. Bush and Mayne, *In (Reluctant) Defense of Enron*, 233–34.

57. The First Circuit had also ruled in *Town of Norwood v. New England Power Co.,* 202 F.3d. 392 (1st Cir.), *cert. denied,* 531 U.S. 818 (2000), that the filed-rate doctrine applied to competitive electricity markets. The application of the antitrust laws to

power markets would have numerous benefits. First, private plaintiffs could seek treble damages. As opposed to single damages, treble damages not only require disgorgement of ill-gotten gains, but also allow for some penalty effect; see 15 U.S.C. § 15. In addition to private plaintiffs, other potential plaintiffs could help enforce the antitrust laws. The U.S. Department of Justice or the Federal Trade Commission could bring suit, and state attorneys general, acting in the capacity of *parens patriae*, could also bring suit under the antitrust laws. See 15 U.S.C. § 15c.

58. 413 F.3d 503 (5th Cir. 2005).

59. TCE cited Tex. Util. Code Ann. § 39.158(b), the statute's savings clause.

60. 413 F.3d at 509.

61. Some independent system operators, such as the New England ISO, use a nodal pricing system to encourage investment in new capacity so as to relieve congestion problems on the grid. Associated Industries of Massachusetts Foundation, *New England's Locational Installed Capacity (LICAP) Market: A Primer* (2005), available at http://www.aimnet.org/AM/Template.cfm?Section=Home&TEMPLATE=/CM/ContentDisplay.cfm&CONTENTID=6367.

62. This process is described well in *Electricity Consumers Resource Council v. Federal Energy Regulatory Commission*, 407 F.3d 1232 (D.C. Cir. 2005). For a description of a similar tactic to ensure that sellers provide adequate reserve margins in New England, see *Central Maine Power Co. v. Federal Energy Regulatory Commission*, 252 F.3d 34 (1st Cir. 2001) and *Sithe New England Holdings LLC v. Federal Energy Regulatory Commission*, 308 F.3d 71 (1st Cir. 2002).

63. The New York independent system operator organizes a process whereby retail sellers acquire their reserve margin capacity at above-market prices that decline as the amount of capacity purchased approaches the target of 118 percent of projected needs. The Midwest ISO has also used capacity auctions.

64. There are two minor exceptions to this. One is FERC's control over siting hydroelectric facilities, which preempts state law; see *First Iowa Hydro-Electric Cooperative v. Federal Power Comm'n*, 328 U.S. 152 (1946). The other is the emergency siting power given FERC under the Energy Policy Act of 2005. See 16 U.S.C. § 824p.

65. See chapter 4.

Chapter 2: Laying the Groundwork for Power Competition in Texas

1. Public Law 95-620, signed by the president on November 9, 1978. U.S. Code, Title 42, Chapter 92.

2. For examples, see *Application of HL&P Co. for Authority to Change Rates*, P.U.C.T. Docket No. 8425, Order on Rehearing (Sept. 18, 1990); *Application of Texas Utilities Electric Company for Authority to Change Rates*, P.U.C.T. Docket No. 11735, Second Order on Rehearing (May 27, 1994).

3. Public Law 95-617, signed by the president on November 9, 1978. U.S. Code, Title 16, Chapter 46.

4. Public Law 74-333, signed by the president on August 26, 1935. U.S. Code, Title 15, Chapter 2C. Each utility operated as the sole provider of integrated generation, transmission, and distribution services in franchise areas. This act required utility holding companies to incorporate in the same state, so they would be subject to state regulation. If the holding company had utilities operating in more than one state, the Securities and Exchange Commission would regulate them.

5. Public Law 95-621, signed by the president on November 9, 1978. U.S. Code, Title 15, Chapter 60.

6. Texas, and in particular the Houston Ship Channel, continues to lead the nation in cogeneration capacity. The data on cogeneration capacity around the United States are collected by the Energy Information Administration through its EIA-860 form and can be found at www.eia.doe.gov/cneaf/electricity/page/eia860.html.

7. Tex-La Electric Cooperative, Inc., F.E.R.C. Docket No. 94-4-000, filed December 15, 1993.

8. Tex-La Electric Cooperative, Inc., 69 FERC ¶ 61,269 (1994).

9. Senate Bill 373, an act relating to the continuation, operations, and functions of the Public Utility Commission of Texas and the Office of Public Utility Counsel; providing penalties. The full text of SB 373 is available at the website of the Texas legislature: http://www.legis.state.tx.us/BillLookup/Text.aspx?LegSess=74R&Bill=SB373.

10. Promoting Wholesale Competition through Open Access Non-Discriminatory Transmission Service by Public Utilities; Recovery of Stranded Costs by Public Utilities and Transmitting Utilities, FERC Order No. 888, F.E.R.C. Stats. & Regs. ¶ 31,036, 75 Fed. Reg. 61,080 (1996) (codified at 18 C.F.R. § 35 (1996)).

11. See, e.g., *Application of Texas Utilities Electric Company for Authority to Implement Rate WPI to Lyntegar Electric Cooperative, Inc. and Taylor Electric Cooperative, Inc.*, Docket No. 14716 (March 21, 1996).

12. Rulemaking on Transmission Pricing and Access (Subst. R. 23.67 & 23.70), P.U.C.T. Docket No. 14045 (1995).

13. Pipeline Service Obligations and Revisions to Regulations Governing Self-Implementing Transportation under Part 284 of the Commission's Regulations and Regulations of Natural Gas Pipelines after Partial Wellhead Decontrol, FERC Order No. 636, 57 Fed. Reg. 13,267 (April 16, 1992), F.E.R.C. Stats. & Regs. Preambles January 1991–June 1996 ¶ 30,939 (April 8, 1992), order on reh'g, Order No. 636-A, 57 Fed. Reg. ¶ 36,128 (August 12, 1992), F.E.R.C. Stats. & Regs. Preambles, January 1991-June 1996, ¶ 30,950 (August 3, 1992), order on reh'g, Order No. 636-B, 57 Fed. Reg. 57,911 (December 8, 1992), 61 FERC ¶ 61,272 (1992), notice of denial of rehearing (January 8, 1993, 62 FERC ¶ 61,007 (1993), aff'd in part and vacated and remanded in part, *UDC v. FERC*, 88 F.3d 1105 (D.C. Cir. 1996), order on remand, Order No. 636-C, 78 FERC ¶ 61,186 (1997), order on reh'g, Order No. 636-D, 83 FERC ¶ 61,210 (1998) (codified at 18 C.F.R. § 284 [1992]).

14. *Application of Central Power & Light for Authority to Change Rates*, Docket No. 14965, Second Order on Rehearing (Oct. 16, 1997).

15. Senate Bill 7, an act relating to electric utility restructuring and to the powers and duties of the Public Utility Commission of Texas, Office of Public Utility Counsel, and Texas Natural Resource Conservation Commission; providing penalties. The full text of SB 7 is available at the website of the Texas legislature: http://www.legis.state.tx.us/BillLookup/Text.aspx?LegSess=76R&Bill=SB7.

16. For example, see Andrew N. Kleit and Timothy Considine, "Can Electricity Restructuring Survive? Lessons from California and Pennsylvania" in *Electric Choices: Deregulation and the Future of Electric Power*, 39–62 (London: Rowman and Littlefield, 2006).

17. Alliance for Retail Markets, Austin, December 2007.

18. Public Utility Commission of Texas, "Scope of Competition in Electric Markets in Texas," Report to Texas Legislature, January 2007. For current prices, offers of competitive REPs can be compared at the official electric choice website of the Public Utility Commission of Texas, "Texas Electric Choice Education Program," www.powertochoose.org.

19. It should be noted that this global settlement did not lead TXU to clean up all of its coal plants at that time.

20. Regional Transmission Organizations, FERC Order No. 2000, 65 Fed. Reg. 809 (Jan. 6, 2000). [1996–2000 Regs. Preambles] F.E.R.C. Stats. & Regs. ¶ 31,089, *order on reh'g*, Order No 2000-A, 65 Fed. Reg. 12,088 (Mar. 8, 2000), [1996–2000 Regs. Preambles] F.E.R.C. Stat. & Regs. ¶ 31,092, *appeal dismissed for want of standing sub nom. Pub. Util. Dist. No.1 v. FERC*, 272 F.3d 607 (D.C. Cir. 2001) (codified at 18 C.F.R. § 35 [2000]).

21. The PUCT had some painful experience with unauthorized provider switches in customer long-distance telephone service through the 1990s and wanted to avoid similar experiences with retail electricity switching. See Order Adopting Amendment to § 26.130, *Relating to Selection of Telecommunications Provider*, P.U.C.T. Project No. 21419 (June 12, 2000). http://interchange.puc.state.tx.us/WebApp/Interchange/Documents/235154.DOC.

22. A website created as part of this education effort still provides an easy way for customers to compare and choose REPs; see Public Utility Commission of Texas, "Texas Electric Choice Education Program," www.powertochoose.org and www.poderdeescoger.org. Unfortunately, statewide budget shortfalls in 2003 cut funding for the programs just when they were needed most, and it was restored only in 2007.

23. For example, see Lehr et al., *Listening to Customers: How Deliberative Polling Helped Build 1,000 MW of New Renewable Energy Projects in Texas* (National Renewable Energy Laboratory, NREL/TP-620-33177, June 2003).

24. Senate Bill 20, relating to this state's goal for renewable energy. The full text of SB 20 is available at the website of the Texas legislature: http://www.legis.state.tx.us/BillLookup/Text.aspx?LegSess=791&Bill=SB20.

25. The data can be found at the American Wind Energy Association's website: www.awea.org/projects/.

26. House Bill 3693, relating to energy demand, energy load, energy-efficiency incentives, energy programs, and energy performance measures. The full text of HB 3693 is available at the website of the Texas legislature: http://www.legis.state.tx.us/BillLookup/Text.aspx?LegSess=80R&Bill=HB3693.

27. Alliance for Retail Markets, Austin, December 2007.

28. Information on retail electricity providers and available products can be found online, with links to individual REP websites provided. See Public Utility Commission of Texas, "Texas Electric Choice Education Program," www.powertochoose.org and www.poderdeescoger.org.

29. Each of these issues is discussed in subsequent chapters in this volume.

30. The historical data can be found at Texas Electricity Profile provided at the website of the EIA: www.eia.doe.gov/cneaf/electricity/st_profiles/texas.html.

31. Annual Energy Review by the EIA provides the historical data on capacity by fuel (table 8.11c). Texas data are tracked by the PUCT: regular updates are provided in New Electric Generating Plants in Texas, which is available at www.puc.state.tx.us/electric/maps/gentable.pdf.

32. Public Utility Commission of Texas, "Scope of Competition in Electric Markets in Texas," Report to Texas Legislature, January 2007. This report is renewed every legislative session (every two years), and current and past issues can be found at http://www.puc.state.tx.us/electric/reports/scope/index.cfm.

33. For example, see Larry Parker and Mark Holt, *Nuclear Power: Outlook for New U.S. Reactors*, Congressional Research Service, Report to Congress, updated March 9, 2007 (http://fas.org/sgp/crs/misc/RL33442.pdf).

34. "NRG Energy Submits Application for New 2,700 Megawatt Nuclear Plant in South Texas." This company announcement can be found at http://www.snl.com/irweblinkx/file.aspx?IID=4057436&FID=4916766.

35. "Tenaska Proposes Nation's First New Conventional Coal-Fueled Power Plant to Capture Carbon Dioxide." This company announcement can be found at http://www.tenaska.com/newsItem.aspx?id=30.

Chapter 3: Evolution of Wholesale Market Design in ERCOT

1. Quoted from the House of Commons Daily Debates (Hansard), November 11, 1947.

2. The elements needed for a sustainable market design are described in Parviz Adib, Eric Schubert, and Shmuel Oren, "Resource Adequacy: Alternate Perspectives and Divergent Paths," in *Competitive Electricity Markets: Design, Implementation and Performance*, ed. Fereidoon P. Sioshansi (London: Elsevier, March 2008).

3. Following the 1994 publication of the "Blue Book" in California to address the restructuring of the industry, the Texas legislature in 1995 opened up wholesale electricity competition and mandated the establishment of a robust transmission network to ensure nondiscriminatory access. The Public Utility Commission of Texas finalized

its decision on a postage-stamp approach to nondiscriminatory access in 1999, when the Texas legislature took the additional step of limiting the ownership of installed-capacity share to 20 percent. California Public Utilities Commission, *Order Instituting Rulemaking and Order Instituting Investigation: On the Commission's Proposed Policies Governing Restructuring California's Electric Services Industry and Reforming Regulation*, R. 94-04-031 and I. 94-04-032, Sacramento, California, April 20, 1994; 76th Session of the Texas Legislature, Senate Bill 7, *An Act Relating to Electric Utility Restructuring; Transmission Service Rates*, P.U.C.T. Substantive Rule 25.192, available at http://puc.state.tx.us/rules/subrules/electric/25.192/25.192.pdf.

4. "Commonwealth markets" such as Australia and New Zealand (and, to a lesser extent, Alberta, Canada), which restructured their generation, retailers, and transmission grids much as ERCOT did, have shown a similar longevity. While these commonwealth markets and ERCOT have taken a variety of approaches to procurement of ancillary services, dispatch of generation units in the real-time market, offer caps, and congestion pricing, they share an emphasis on lighthanded regulation, with strong reliance on retail competition and treatment of electricity as a commodity.

5. In a 2006 survey, the ERCOT retail market was ranked the fourth most active in the world, based on retail switching. The three higher-ranked markets were the United Kingdom and the Australian states of Victoria and South Australia. All four are underpinned by wholesale markets that use an energy-only resource adequacy mechanism. Paul Grey, "Texas Is the Rising Star," *Public Utilities Fortnightly's Spark*, Letter No. 31, July 2006, available at http://www.pur.com/pubs/spark/jul06.pdf (accessed February 11, 2009). See chapter 4, below, for a discussion on energy-only resource adequacy mechanisms. See also Parviz Adib and Jay Zarnikau, "Texas: The Most Robust Competitive Market in North America," in *Electricity Market Reform: An International Perspective*, ed. Fereidoon P. Sioshansi and Wolfgang Pfaffenberger (Amsterdam: Elsevier, 2006).

6. It is also helpful to mention that these markets, particularly Pennsylvania–New Jersey–Maryland (PJM), the New York independent system operator (NYISO), and the New England independent system operator (ISO-NE), had been operated as power pools before their restructuring, when economic dispatch was a norm in their operations. Their market design choice was, therefore, consistent with their historical operations.

7. The troubled evolution of the capacity market approach was highlighted in detail in Adib et al., "Resource Adequacy."

8. Energy Retailer Research Consortium, *Annual Baseline Assessment of Choice in Canada and the United States, Residential* (Washington, D.C.: Energy Retailer Research Consortium, 2008).

9. See chapters 4 and 5, below, for more on this point.

10. Stakeholders—that is, participants in ERCOT committee and subcommittee meetings—included competitive retailers, independent power producers, incumbent utilities, electric cooperatives, municipally owned utilities, transmission and distribution

providers, power marketers, industrial consumers, and consumer advocacy groups.

11. In September 2002, the Market Oversight Division of the PUCT identified various shortcomings associated with the operation of a zonal design in the ERCOT market. Moving to a nodal design was among the remedies identified; it has taken a significant amount of time and effort, however, to complete this move, currently scheduled for December 2010.

12. "MOD" refers throughout this chapter to the division's staff and its senior advisor, Dr. Shmuel Oren. At ERCOT stakeholder meetings, MOD served two different roles. First, it represented the commissioners in discussions involving the rules and orders they had signed. Second, it represented the public interest on issues that were not directly affected by these rules and orders. In addition to Parviz Adib and Eric Schubert, MOD members included Danielle Jaussaud, Richard Greffe, Julie Gauldin, Tony Grasso, Teresa Kirk, David Hurlbut, and Sam Zhou. MOD was also privileged to work with PUCT staff attorney Keith Rogas on market design issues.

13. For a more complete discussion of the impact of a single regulator on the development of the ERCOT market, see the conclusion to chapter 4, below.

14. Parviz Adib was appointed as MOD's director, and Eric Schubert was one of the division's original senior economists.

15. For related market design rules, see P.U.C.T. Subst.R. §§ 25.501 through 25.505, available at http://www.puc.state.tx.us/rules/subrules/electric/index.cfm.

16. In a centralized pool, the owners of each online generation unit offer the unit's entire potential output for centralized dispatch by the system operator. In the ERCOT zonal model, owners of each online generation unit were not required to have any of its output centrally dispatched by the system operator except when the reliability of the grid was at risk.

17. Wind farms did not become an important part of the generation-siting process until after the real-time energy market began on July 31, 2001.

18. A QSE is a market participant that is qualified by ERCOT to submit energy offers, energy schedules, and ancillary services bids, as well as settle payments with ERCOT.

19. A load is an end-user of electricity. Load-serving entities in ERCOT are competitive retailers, municipally owned utilities, and electric cooperatives.

20. Shmuel Oren, "Report to the Public Utility Commission of Texas on the ERCOT Protocols," *Petition of the Electric Reliability Council of Texas for Approval of the ERCOT Protocols*, P.U.C.T. Docket No. 23220 (February 9, 2001), 5–6.

21. See ERCOT Market Protocols § 7.4.

22. "Order on Rehearing," *Petition of the Electric Reliability Council of Texas (ERCOT) for Approval of the ERCOT Protocols*, P.U.C.T. Docket No. 23220 (June 4, 2001).

23. Market Oversight Division, "Comments on Issues Related to Transmission Congestion Workshop on September 18, 2002," *Transmission Congestion Issues in the Electric Reliability Council of Texas*, P.U.C.T. Project No. 26376 (September 2002). Available

at http://interchange.puc.state.tx.us/WebApp/Interchange/application/dbapps/filings/pgSearch_Results.asp?TXT_CNTR_NO=26376&TXT_ITEM_NO=19.

24. Eric Schubert, "An Energy-Only Resource Adequacy Mechanism," P.U.C.T. Rulemaking Project No. 24255, *Rulemaking Concerning Planning Reserve Margin Requirements* (April 14, 2005), available at http://interchange.puc.state.tx.us/WebApp/Interchange/Documents/24255_98_475491.PDF (accessed February 11, 2009).

25. Market Oversight Division, "Comments on Issues Related to Transmission Congestion Workshop," 5.

26. Resource-specific offer curves refer to a system where each resource provides ERCOT operators with an offer curve for dispatch in the real-time market. Under the ERCOT zonal market, generation owners are only required to provide a portfolio offer curve where the individual resources of a particular generation owner are aggregated by zone and placed into one curve, giving the generation owner freedom to choose which of these resources to dispatch in real time.

27. Order adopting P.U.C.T. Substantive Rule § 25.501, *Wholesale Market Design for the Electric Reliability Council of Texas*, P.U.C.T. Project No. 26376 (Sept. 22, 2003), available at http://puc.state.tx.us/rules/rulemake/26376/26376adt1.pdf; Order on Rehearing at 19, *Petition of the Electric Reliability Council of Texas for Approval of the ERCOT Protocols*, P.U.C.T. Docket No. 23220 (June 4, 2001).

28. For a good discussion on this issue, see Peter Cramton, "Electricity Market Design: The Good, the Bad, and the Ugly," in *Proceedings of the Hawaii International Conference on System Sciences*, Shidler College of Business, University of Hawaii at Manoa, January 2003.

29. Independent power producers and out-of-state generators, with experience in northeastern U.S. nodal markets, were among the few parties supporting a transition to nodal market operation.

30. Retailers forcefully made these arguments during the subsequent resource adequacy rulemaking, PUCT Rulemaking Project No. 31972. During the development of the nodal market protocols, independent power producers, like industrial customers, assumed that a nodal market design would inevitably lead to ICAP markets.

31. See chapter 4, below. The New Zealand market combines nodal dispatch with an energy-only market, while the Australian market, which is zonal, has a real-time market that is a mandatory pool (similar to nodal) rather than the physical bilateral market on which the ERCOT zonal market was based. Non-U.S. market designs were not well known among ERCOT stakeholders at this time.

32. See chapter 4, below, for a more detailed discussion of this topic; also see Adib et al., "Resource Adequacy."

33. The Texas-based owners of large portfolios were TXU, Texas Genco (now NRG), Austin Energy, City Public Service of San Antonio, and the Lower Colorado River Authority (LCRA). Independent power producers included American National Power, Calpine, Constellation Power, Dynegy Inc., El Paso Corporation, FPL Energy, PG&E National Energy Group, TECO Energy, and Texas Independent Energy.

34. These limitations included ramp rates (that is, the speed at which a generation unit can increase or decrease output), minimum loadings (the lowest level of output that a unit can safely or economically maintain), and multiple configurations of units (varying combinations of gas-fired turbines and steam units that comprise combined-cycle units).

35. Order Adopting New § 25.501, *Wholesale Market Design for the Electric Reliability Council of Texas*, P.U.C.T. Project No. 26376 (Sept. 22, 2003), available at http://puc.state.tx.us/rules/rulemake/26376/26376adt1.pdf.

36. Non-opt-in entities comprised all Texas municipalities and cooperatives within ERCOT, who were allowed by the 1999 Senate Bill 7 to opt out of retail competition.

37. Tabors Caramanis & Associates and KEMA Consulting, *Market Restructuring Cost Benefit Analysis*, Final Report, November 30, 2004, available at http://interchange.puc.state.tx.us/WebApp/Interchange/Documents/28500_28_465104.PDF (accessed February 11, 2009).

38. In 2008, the PUCT requested an update to the cost-benefit analysis. See CRA International, "Update on the ERCOT Nodal Market Cost-Benefit Analysis," *Transition to an ERCOT Nodal Market Design*, P.U.C.T. Project No. 31600 (December 18, 2008). The updated results confirmed the finding of the initial analysis that the move to a nodal market design was cost-effective.

39. Texas Nodal Team, ERCOT, "Comments on Texas Nodal Team White Papers" by Shmuel Oren (April 24, 2004; revised May 18, 2004); "Comments on Texas Nodal Team White Papers" by David Patton (June 24, 2004); and "Comments on Texas Nodal Team White Papers" by Frank Wolak (May 18, 2004). All three documents available at http://nodal.ercot.com/docs/tntarc/er/; Roy Shanker, "Comments on Texas Nodal Team Market Design Proposals" (August 2004). Available at http://nodal.ercot.com/docs/tntarc/er/ShankerTexasNodalReview082304.DOC.

40. Tabors Caramanis & Associates and KEMA Consulting, Inc., "Market Restructuring Cost-Benefit Analysis," *Activities Related to the Implementation of a Nodal Market for the Electric Reliability Council of Texas*, P.U.C.T. Docket No. 28500 (December 21, 2004). For the full text, please see: http://interchange.puc.state.tx.us/WebApp/Interchange/application/dbapps/filings/pgSearch_Results.asp?TXT_CNTR_NO=28500&TXT_ITEM_NO=28.

41. During the day-ahead reliability unit commitment (RUC) process, ERCOT selects certain resources to be available during certain settlement intervals. It guarantees that those resources are reimbursed for the costs to start up and operate the unit at a minimum sustainable level to ensure that the resource owner does not lose money when providing reliability service that ERCOT requested. If, however, the actual real-time operation results in revenues in excess of the amount guaranteed by ERCOT, a relatively high RUC "clawback" charge is assessed to the resource owner in question. This provision encourages resource owners to schedule their resources the day before the operating day if they expect them to function economically during the operating day.

42. Testimony of Dr. David Patton, *Proceeding to Consider Protocols to Implement a Nodal Market in the Electric Reliability Council of Texas (ERCOT) Pursuant to P.U.C. Subst. R. § 25.501*, P.U.C.T. Docket No. 31540 (November 11, 2005). For direct testimony, see http://interchange.puc.state.tx.us/WebApp/Interchange/application/dbapps/filings/pgSearch_Results.asp?TXT_CNTR_NO=31540&TXT_ITEM_NO=124. For revisions to direct testimony, see http://interchange.puc.state.tx.us/WebApp/Interchange/Documents/31540_149_496724.PDF.

43. The twenty-one-page final decision by the commission was filed in April 2006. "Final Order *Proceeding to Consider Protocols to Implement a Nodal Market in the Electric Reliability Council of Texas Pursuant to Subst. R. §25.501*," P.U.C.T. Docket No. 31540 (April 5, 2006). Available at http://interchange.puc.state.tx.us/WebApp/Interchange/application/dbapps/filings/pgSearch_Results.asp?TXT_CNTR_NO=31540&TXT_ITEM_NO=303.

44. Electric Reliability Council of Texas, *Report on Existing and Potential Electric System Constraints and Needs*, December 2006, 27, available at http://www.ercot.com/news/presentations/2006/2006_ERCOT_Reports_Transmission_Constraints_and_Needs.pdf (accessed February 11, 2009).

45. Another benefit of zonal energy prices for loads was that they greatly facilitated mass market retailing, an important concern for the ERCOT market. Order Adopting New P.U.C.T. Substantive Rule § 25.501, *Wholesale Market Design for the Electric Reliability Council of Texas*, P.U.C.T. Project No. 26376 (Sept. 22, 2003), 123–24, available at http://puc.state.tx.us/rules/rulemake/26376/26376adt1.pdf.

46. The three PUCT-approved NOIE load zones are the City of Austin, the City of San Antonio, and the Lower Colorado River Authority. Several other NOIEs are currently evaluating the benefits of requesting their own load zones.

47. It is highly possible that most of the costs imposed on generators in this order will eventually be reflected in prices charged to end-use customers, as often happens with government and regulatory taxes and fees.

48. Final Order *Proceeding to Consider Protocols to Implement a Nodal Market in the Electric Reliability Council of Texas Pursuant to Subst. R. §25.501*," P.U.C.T. Docket No. 31540 (April 5, 2006). Available at http://interchange.puc.state.tx.us/WebApp/Interchange/application/dbapps/filings/pgSearch_Results.asp?TXT_CNTR_NO=31540&TXT_ITEM_NO=303.

49. See Michael T. Burr, "Razing the Regulatory Compact: Interview with Lynne Kiesling," *Public Utilities Fortnightly* 145, no. 9 (September 2007): 18–20, 82.

50. Examples of rapid adoptions of innovations in ERCOT that could be considered investment bubbles are the "wind rushes" in 2001 and 2007–8 and the combined-cycle bubble of 1999–2003. Innovations in financial markets that have occurred at times over the past three centuries are closely analogous to these investment bubbles. For a historical perspective of the evolution of financial markets and the occasional disruptions that financial innovations cause, see Charles P. Kindleberger, *Manias, Panics, and Crashes: A History of Financial Crises*, 4th ed. (New York: John Wiley and Sons, 2000).

Chapter 4: Achieving Resource Adequacy in Texas via an Energy-Only Electricity Market

1. Nordpool is an electricity market with a footprint that includes Denmark, Finland, Norway, and Sweden.

2. There are regions within Texas that are not part of ERCOT, where the Federal Energy Regulatory Commission (FERC) rather than the Public Utility Commission of Texas, exerts jurisdiction on wholesale market issues.

3. See, for example, Joseph E. Bowering, "The Evolution of PJM's Capacity Market," in *Competitive Electricity Markets: Design, Implementation and Performance*, ed. Fereidoon P. Sioshansi, 363–86 (London: Elsevier, 2008).

4. Offer caps are common in various competitive electricity markets where offer prices are required not to exceed certain thresholds. The most popular offer-cap threshold in U.S. electricity markets is $1,000 per MWh or per MW per hour. As of this writing, the ERCOT electricity market has an offer cap of $2,250 per MWh, which is scheduled to be raised to $3,000 per MWh in early 2011 after the opening of the new nodal market design.

5. See, for example, B. F. Hobbs, J. Inon, and S. E. Stoft, "Installed Capacity Requirements and Price Caps: Oil on the Water, or Fuel on the Fire?" *Electricity Journal* 14 (June 2001): 23–34.

6. See, for example, M. Abbott, "Is the Security of Electricity Supply a Public Good?" *Electricity Journal* 14 (July 2001): 31–33.

7 S. S. Oren, "Ensuring Generation Adequacy in Competitive Electricity Markets," in *Electricity Deregulation: Choices and Challenges, Bush School Series in the Economics of Public Policy*, ed. James M. Griffin and Steven L. Puller (Chicago: University of Chicago Press, 2005).

8. See ibid.

9. See Lynne Kiesling, "The Role of Retail Prices in Electricity Restructuring," in *Electric Choices: Deregulation and the Future of Electric Power*, ed. Andrew N. Kleit, 39–62 (Rowman and Littlefield, 2007).

10. PURA § 39.108, *Financial Standards for Retail Electric Providers Regarding the Billing and Collection of Transition Charges*.

11. See chapter 2. Seventy-sixth Session of the Texas Legislature, Senate Bill 7, *An Act Relating to Electric Utility Restructuring*.

12. See chapter 5.

13. Paul Grey, "Texas Is the Rising Star," *Public Utilities Fortnightly's Spark*, Letter No. 31, July 2006, available at http://pur.cs.net/pubs/spark/jul06.pdf (accessed February 11, 2009).

14. Public Utility Commission of Texas, *New Electric Generation Plants in Texas*, April 23, 2009, http://www.puc.state.tx.us/electric/maps/gen_tables.xls (accessed June 14, 2009).

15. The troubled evolution of the capacity-market approach has been highlighted

in detail in Parviz Adib, Eric Schubert, and Shmuel Oren, "Resource Adequacy: Alternate Perspectives and Divergent Paths" in *Competitive Electricity Markets: Design, Implementation and Performance*, ed. Fereidoon P. Sioshansi (London: Elsevier, 2008).

16. See note 15, above.

17. See chapter 3.

18. The zonal design used by NEMMCO included a mandatory centralized pool, with side arrangements to address pockets of local congestion. This approach was inconsistent with the desires of participants who preferred to use the zonal market to clear only the real-time energy imbalances to physical bilateral transactions in the wholesale energy market.

19. The Energy Intermarket Surveillance Group (EISG) is a voluntary organization that includes about twenty market monitors from all electricity markets in the United States and Canada, as well as monitors from other electricity markets throughout the world.

20. Eric Schubert, "An Energy-Only Resource Adequacy Mechanism," *Rulemaking Concerning Planning Reserve Margin Requirements,* P.U.C.T. Rulemaking Project No. 24255 (April 14, 2005), available at http://interchange.puc.state.tx.us/WebApp/Interchange/Documents/24255_98_475491.PDF (accessed February 11, 2009).

21. Barry T. Smitherman, Commissioner, Memorandum, *Rulemaking Concerning Planning Reserve Margin Requirements,* P.U.C.T. Rulemaking Project No. 24255 (July 15, 2005), available at http://interchange.puc.state.tx.us/WebApp/InterchangeDocuments/24255_175_484093.pdf.

22. Order Adopting Amendment to Substantive Rule § 25.502, New Substantive Rule § 25.504, and New Substantive Rule § 25.505, *Rulemaking on Wholesale Electric Market Power and Resource Adequacy in the ERCOT Power Region,* P.U.C.T. Project No. 31972 (August 24, 2006), 6, available at http://interchange.puc.state.tx.us/WebApp/Interchange/application/dbapps/filings/pgSearch_Results.asp?TXT_CNTR_NO=31972&TXT_ITEM_NO=78 (accessed February 11, 2009).

23. Such a contracting requirement for competitive retailers also would have obliged owners of generation that received contracts from competitive retailers to offer a specified amount of the generation into the real-time market.

24. The ERCOT, Australian, New Zealand, and Alberta markets, all having energy-only resource adequacy mechanisms, have financially binding scheduling near real time. All nodal markets in the United States have, in addition, a financially binding day-ahead market, also known as a "two-settlement" market.

25. The following discussion has been taken from Eric Schubert, David Hurlbut, Shmuel Oren, and Parviz Adib, "The Texas Energy-Only Resource Adequacy Mechanism," *Electricity Journal* 19 (December 2006): 39–49.

26. David Hurlbut and Eric S. Schubert, Memorandum, *Rulemaking Concerning Planning Reserve Requirements*, P.U.C.T. Project No. 24255 (October 31, 2005), available at http://interchange.puc.state.tx.us/WebApp/Interchange/Documents/24255_201_494897.PDF (accessed February 11, 2009).

27. Order Adopting Amendment to Substantive Rule § 25.502, 146.

28. Ibid., 42.

29. Order Adopting New § 25.130, *Rulemaking Related to Advanced Metering*, P.U.C.T. Project No. 31418 (May 10, 2007), available at http://interchange.puc.state.tx.us/WebApp/Interchange/application/dbapps/filings/pgSearch_Results.asp?TXT_CNTR_NO=31418&TXT_ITEM_NO=110 (accessed February 11, 2009).

30. *Oncor Electric Delivery Company LLC's Request for Approval of Advanced Metering System (AMS) Deployment Plan and Request for AMS Surcharge*, P.U.C.T. Docket No. 35718, Final Order (August 22, 2008), available at http://interchange.puc.state.tx.us/WebApp/Interchange/application/dbapps/filings/pgSearch_Results.asp?TXT_CNTR_NO=35718&TXT_ITEM_NO=102 (accessed February 11, 2009); *Application of Centerpoint Energy Houston Electric, LLC for Approval of Deployment Plan and Request for Surcharge for an Advanced Metering System*, P.U.C.T. Docket No. 35639, Final Order (December 22, 2008), available at http://interchange.puc.state.tx.us/WebApp/Interchange/application/dbapps/filings/pgSearch_Results.asp?TXT_CNTR_NO=35639&TXT_ITEM_NO=219 (accessed February 11, 2009).

31. Order Adopting Amendment to Substantive Rule § 25.502, 27–28.

32. *Constellation Energy Commodities Group, Inc. v. Public Util. Comm'n*, No. 03-06-00552-CV (Tex. App. Austin, filed September 14, 2006), and *City of Garland v. Public Util. Comm'n*, No. 03-06-00571-CV (Tex. App. Austin, filed September 21, 2006).

33. Order Adopting Amendment to § 25.502, *Pricing Safeguards in Markets Operated by the Electric Reliability Council of Texas*, P.U.C.T. Project No. 33490, *Rulemaking Proceeding to Amend §25.502; Pricing Safeguards in Markets Operated by the Electric Reliability Council of Texas* (August 16, 2007). Available at http://interchange.puc.state.tx.us/WebApp/Interchange/application/dbapps/filings/pgSearch_Results.asp?TXT_CNTR_NO=33490&TXT_ITEM_NO=24 (accessed February 11, 2009).

34. For a full copy of *Resource Adequacy in the Electric Reliability Council of Texas Power Region*, P.U.C.T. Substantive Rule § 25.505, available at http://www.puc.state.tx.us/rules/subrules/electric/25.505/25.505.pdf (accessed February 11, 2009).

35. As P.U.C.T. Substantive Rule § 25.505(g)(6) states, "The low system offer cap shall be set on a daily basis at the higher of: (i) $500 per MWh and $500 per MW per hour; or (ii) 50 times the daily Houston Ship Channel gas price index of the previous business day, expressed in dollars per MWh and dollars per MW per hour," available at http://www.puc.state.tx.us/rules/subrules/electric/25.505/25.505.pdf (accessed February 11, 2009).

36. Alan Moran and Ben Skinner, "Resource Adequacy and Efficient Infrastructure Investment," in *Competitive Electricity Markets: Design, Implementation and Performance*, ed. by Fereidoon P. Sioshansi (London: Elsevier, 2008), 394, figure 11.3.

37. See, for example, W. Hogan, "On an 'Energy Only' Electricity Market Design for Resource Adequacy," presented at the Eleventh Annual POWER Research Conference on Electricity Regulation and Restructuring, Berkeley, Calif., March 2006, and

P. Joskow and J. Tyrol, "Reliability and Competitive Electricity Markets," *RAND Journal of Economics* 38-1 (Spring 2007): 60–80.

38. Potomac Economics, LTD, "Report on the Wholesale Market Events of March 3, 2008," *PUC Market Oversight Activities*, P.U.C.T. Project No. 23100, March 24, 2008. Available at http://interchange.puc.state.tx.us/WebApp/Interchange/application/dbapps/filings/pgSearch_Results.asp?TXT_CNTR_NO=23100&TXT_ITEM_NO=47 (accessed February 11, 2009). At the time of this writing, at the suggestion of the ERCOT Independent Market Monitor (IMM), ERCOT Protocol Revision Request (PRR) No. 791, *Scarcity Pricing Mechanism*, was being considered by ERCOT stakeholders to provide an administrative pricing mechanism to supplement the "small fish swim free" market-mitigation approach. The authors anticipate that the combination of the two approaches will provide more consistent scarcity pricing for the ERCOT market without the potential problems associated with physical withholding.

39. Given the existence of such programs as Genscape, the PUCT staff assumed that the larger and more sophisticated players generally knew the bidding strategies of their competitors.

40. See, for example, Edward J. Green and Robert H. Porter, "Noncooperative Collusion under Imperfect Price Information," *Econometrica* 52, no. 1 (January 1984): 87–100, and G. Stigler, "A Theory of Oligopoly," *Journal of Political Economy* 72 (1964): 44–61.

41. Australian market monitors to PUCT staff, private electronic correspondence, spring 2006.

42. In addition to communicating with the Australian market monitor, Peter Adams, the commission staff contacted several academicians, such as professors Shmuel Oren, Frank Wolak, and Steve Puller, who felt that premature information disclosure could harm the market. The staff was also influenced by positions taken by the U.S. Department of Justice and the U.S. Federal Trade Commission that were supportive of delay in disclosure of information well beyond the two days in action in the Australian market.

43. Luminant (formerly TXU Generation Company) filed the first voluntary mitigation plan that was approved by the commission in August 2007. Please see P.U.C.T. Docket No. 34480, *TXU Wholesale Companies' Request for Approval of a Voluntary Mitigation Plan Pursuant to P.U.C. Substantive Rule 25.504(e)*. For a critique of this mitigation plan, see chapter 9, below.

44. F. Wolak, "An Empirical Analysis of the Impact of Hedge Contracts on Bidding Behavior in a Competitive Electricity Market," *International Economic Journal* 14, no. 2 (Summer 2000): 1–39.

45. Dan Jones, "Wholesale Market Update," presentation to the ERCOT Technical Advisory Committee, February 7, 2008, available at http://www.ercot.com/meetings/tac/keydocs/2008/0207/15._20070207_TAC_Meeting_%28D_Jones%29.ppt (accessed February 11, 2009).

46. *Electric Reliability Council of Texas (ERCOT) Emergency Interruptible Load Service (EILS)*, P.U.C.T. Subst. R. § 25.507, available at http://www.puc.state.tx.us/rules/subrules/electric/25.507/25.507.pdf (accessed February 11, 2009).

47. Under the upcoming nodal market design, a greater range of controllable loads will be able to participate in the operating reserves markets on par with generation resources and curtail their electricity use in response to real-time energy prices. In addition, the PUCT may allow residential and small commercial loads with advanced meters to be settled on their actual usage rather than on a historical average profile. The benefits of allowing all loads to be settled on their actual usage with respect to grid reliability and resource adequacy could be substantial. For a more complete discussion, see BP Energy Company, "Comments on Questions Related to Implementing 15-Minute Settlement for Residential and Small Commercial Customers with Advanced Meters," *Implementation Project Related to Advanced Metering,* P.U.C.T. Project No. 34610 (March 28, 2008), available at http://interchange.puc.state.tx.us/WebApp/Interchange/Documents/34610_42_579661.PDF (accessed February 11, 2009).

48. While still too soon to give all the credit to a decision in 2006 by the PUCT to go with an energy-only electricity market, the latest report on the capacity, demand, and reserves in the ERCOT region, published on May 29, 2009, shows adequate reserves through 2014. For the full report, see http://www.ercot.com/content/news/presentations/2009/2009_ERCOT_Capacity,_Demand_and_Reserves_Report.pdf (accessed June 14, 2009).

49. Two decisions by the ERCOT Board of Directors in June 2008 addressed problems that resulted in these unusual price spikes. First, the approval of Protocol Revision Request 764 (PRR764) addressed some of the inefficiencies in the ERCOT congestion management procedures and resulted in better determination of transmission-limiting elements. Second, as requested by the commission, the board decided to impose a price cap equal to the existing offer cap for the first time since the ERCOT market opened for retail competition on July 31, 2001. Minutes of the meeting are available at http://www.ercot.com/content/meetings/board/keydocs/2008/0715/Item_3a_-_2008_06_06_DRAFT_Board_Meeting_Minutes.pdf (accessed February 11, 2009).

50. At this writing, MISO is proposing a pricing mechanism that would allow energy prices to rise to $3,500 when the energy price hits its cap of $1000/MWh and the reserve penalty factor, which is set administratively through a demand function for operating reserves, reaches its $2,500/MWh maximum. The MISO resource adequacy proposal was approved in October 2008 by FERC in FERC Docket No. ER08-394. Available at http://www.ferc.gov/whats-new/comm-meet/2008/101608/E-8.pdf (accessed February 11, 2009).

51. See chapter 5.

Chapter 5: Texas Transmission Policy

1. Texas policy on competition is set out in Public Utility Regulatory Act, Tex. Util. Code § 39.001, et seq. (2007), available at www.puc.state.tx.us/rules/statutes/index.cfm.

2. These integrated utilities existed when the open-access reforms were completed in Texas, but, as a result of restructuring associated with retail competition, and mergers and acquisitions, they no longer exist.

3. The cogeneration and renewable energy provisions were in the Public Utility Regulatory Policies Act of 1978, Pub. L. 95-617, Nov. 9, 1978, 92 Stat. 3117 (16 U.S.C. 2601 et seq.); relevant provisions codified at 16 USC §824a-3; transmission provisions of the Energy Policy Act of 1992 (Pub. L. 102-486, Oct. 24, 1992, 106 Stat. 2776) are codified at 16 USC §824j.

4. P.U.C.T. Subst. Rule § 23.67, 21 Tex. Reg. 1397 (1996), available at http://tex-info.library.unt.edu/texasregister/1996.htm; Order No. 888, Promoting Wholesale Competition through Open Access Non-Discriminatory Transmission Service by Public Utilities; Recovery of Stranded Costs by Public Utilities and Transmitting Utilities, F.E.R.C. Stats. & Regs. ¶ 31,036, 61 Fed. Reg. 21,540 (1996).

5. FERC's authority is over interstate wholesale sales of electricity. Wholesale sales within the ERCOT region are regarded as intrastate and were subject to PUCT regulation prior to the introduction of wholesale competition. See chapter 1.

6. Public Utility Regulatory Act, Tex. Util. Code § 39.151, available at www.puc.state.tx.us/rules/statutes/index.cfm; current PUCT rules are Substantive Rules §§ 25.191–25.230, 16 Texas Admin. Code §§ 25.191–25.203, available at www.puc.state.tx.us/rules/subrules/electric/index.cfm.

7. *Standardization of Generator Interconnection Agreements and Procedures,* Order No. 2003, 104 F.E.R.C. ¶ 61,103 (2003), 68 Fed. Reg. 49845 (Aug. 19, 2003).

8. P.U.C.T. Subst. R. § 25.195(c)(1), 16 Texas Admin. Code §25.195(c)(1), available at www.puc.state.tx.us/rules/subrules/electric/index.cfm.

9. *Open Access Same-Time Information System (Formerly Real-Time Information Networks) and Standards of Conduct*, Order No. 889, 83 F.E.R.C. ¶ 61,301 (1998), 63 Fed. Reg. 54257 (Oct. 8, 1998).

10. *See Preventing Undue Discrimination and Preference in Transmission Service,* Order No. 890, 118 F.E.R.C. ¶ 61,119 (2007), 72 Fed. Reg. 12265 (Mar. 15, 2007). This order directs utilities and regional transmission organizations to reform the transmission tariffs but does not adopt the broad reforms some parties have advocated. The order notes, "Occidental [Chemical Corporation] claims that it has first-hand experience with a vertically integrated transmission provider that, despite having an OATT, appears to have persistently used its transmission system to preferentially benefit its merchant function." Similar comments have come from the Williams Power Company.

11. See Public Utility Commission of Texas, New Electric Generating Plants in

Texas (Sept. 2, 2008), at www.puc.state.tx.us/electric/maps/gentable.pdf. (accessed March 13, 2009).

12. *Remedying Undue Discrimination through Open Access Transmission Service and Standard Electric Market Design*, FERC RM01-12-000. This proposal was withdrawn, but additional reforms were adopted in 2007 to increase the transparency of transmission service rules. See *Preventing Undue Discrimination and Preference in Transmission Service,* Order No. 890, 118 F.E.R.C. at ¶ 61,119.

13. P.U.C.T. Subst. Rule § 23.67, 21 Tex. Reg. 1397 (1996).

14. Ibid; *Regional Transmission Proceeding to Establish Postage Stamp Rate and Statewide Loadflow Pursuant to Subst. Rule § 23.67*, P.U.C.T. Docket No. 15840, Order on Rehearing (Oct. 3, 1997).

15. No federal entities owned substantial electrical facilities in the ERCOT region.

16. The development of regional transmission organizations (RTOs) has resulted in regional assessments of the availability of transmission service by an RTO.

17. Among the reforms adopted by FERC in 2007 is the requirement that transmission operators adopt an open, transparent, and coordinated planning process. *Preventing Undue Discrimination and Preference in Transmission Service,* Order No. 890, 118 F.E.R.C. at ¶ 61,119.

18. Public Utility Regulatory Act, Tex. Util. Code § 37.056, available at www.puc.state.tx.us/rules/statutes/index.cfm.

19. The ERCOT planning process is summarized in ISO/RTO Planning Committee, *ISO/RTO Electric System Planning: Current Practices, Expansion Plans and Planning Issues*, 2006, available at www.ercot.com/news/presentations/2006/IRC_PC_Planning_Report_Final_02_06_06.pdf (accessed February 11, 2009).

20. Electric Reliability Council of Texas, *2006 Annual Report of the Electric Reliability Council of Texas*, www.ercot.com/news/presentations/2007/2006_Annual_Report.pdf (accessed February 11, 2009).

21. Public Utility Commission of Texas, Summary of Non-bypassable Charges for TDU's beginning January 1, 2002 (Oct. 23, 2008), at www.puc.state.tx.us/electric/rates/Trans/TDGenericRateSummary.pdf (accessed March 13, 2009).

22. P.U.C.T. Subst. R. § 25.192(g), 16 Texas Admin. Code § 25.192(g).

23. For example, TXU Electric Delivery's application to increase its transmission rates to reflect additional transmission capital investment of about $193 million, filed on February 28, 2005, was approved on May 4, 2005. *Application of TXU Electric Delivery Company for Interim Update of Wholesale Transmission Rates Pursuant to Substantive Rule 25.192(G)(1),* P.U.C.T. Docket No. 30802 (May 4, 2005).

24. P.U.C.T. Subst. R. § 25.193, 16 Texas Admin. Code § 25.193. Until recently, FERC-regulated utilities operating in areas of Texas outside of ERCOT did not have the ability to recover increases in transmission charges from retail customers, except through a rate case. Legislation was adopted in the 2007 session to allow expedited adjustments for utilities in the Southwest Power Pool or Western Electricity

Coordinating Council. Public Utility Regulatory Act, Tex. Util. Code § 36.209, available at www.puc.state.tx.us/rules/statutes/index.cfm.

25. Electric Reliability Council of Texas, *2006 Annual Report*.

26. Public Utility Regulatory Act, Tex. Util. Code § 39.151, available at www.puc.state.tx.us/rules/statutes/index.cfm.

27. See chapter 9, below.

28. As of the end of 2008, the independent members of the ERCOT board were chairman Jan Newton, vice chairman Michehl Gent, Mark G. Armentrout, Miguel Espinosa, and Alton D. Patton. The customer representatives were Andrew J. Dalton (industrial), Nick Fehrenbach (large commercial), and Don Ballard (residential and small commercial). The sector representatives were Brad Cox (power marketers), Bob Helton (nonutility generators), Charles Jenkins (investor-owned utilities), Clifton Karnei (electric cooperatives), Robert Thomas (retail electric providers), and Dan Wilkerson (municipally owned utilities). Barry Smitherman, PUCT chairman, served as a nonvoting member, and the chief executive officer was Bob Kahn.

29. P.U.C.T. Subst. Rule § 25.362, 16 Texas Admin. Code § 25.362.

30. Order Adopting Amendments to Substantive. Rule § 25.361 and New Substantive Rule § 25.362 as Approved at the February 13, 2003 Open Meeting, *Rulemaking on Oversight of Independent Organizations in the Competitive Electric Market*, P.U.C.T. Project No. 25959 (Mar. 5, 2003).

31. Greg Abbott (Texas attorney general), "Update on the ERCOT Corruption Case," Weekly AG Newspaper Columns (September 2006), http://www.oag.state.tx.us/newspubs/weeklyag/2006/0906ercot.pdf (accessed February 11, 2009).

32. Michael Dworkin and Rachel Goldwasser, *Ensuring Consideration of the Public Interest in the Governance and Accountability of Regional Transmission Organizations*, Energy Law J. Vol. 28, No. 2 (2007): 543–601.

33. The fact that the RTOs are nonprofit organizations also makes it difficult for regulatory bodies to design effective sanctions in response to poor performance.

34. *Application of the Electric Reliability Council of Texas for Approval of the ERCOT System Administration Fee*, P.U.C.T. Docket No. 31824, Final Order (2006), available at http://interchange.puc.state.tx.us/WebApp/Interchange/application/dbapps/filings/pgSearch.asp.

35. In the report on the August 2003 blackout in the northeastern United States and southeast Canada prepared by the staffs of Natural Resources Canada and the U.S. Department of Energy, the authors identified deficiencies in procedures and tools of the Midwest Independent System Operator (MISO) and in communications procedures between MISO and the PJM Interconnection as causes of the blackout. U.S.–Canada Power System Outage Task Force, *Final Report on the August 14, 2003 Blackout in the United States and Canada: Causes and Recommendations*, April 2004, available at https://reports.energy.gov/ (accessed March 13. 2009).

36. Craig Pirrong, "Transactions Costs and the Organization of Coordination Activities in Power Markets," in *Electric Choices*, ed. Andrew N. Kleit, 113–34 (Lanham, MD: Rowman and Littlefield, 2007).

Chapter 6: Distributed Generation Drives Competitive Energy Services in Texas

1. A detailed presentation of the benefits of distributed electrical resources appears in Amory B. Lovins, E. Kyle Datta, Thomas Feiler, Karl R. Rábago, Joel N. Swisher, André Lehmann, and Ken Wicker, *Small Is Profitable: The Hidden Economics of Making Electrical Resources the Right Size* (Snowmass, CO: Rocky Mountain Institute, 2002).

2. Pub. L. 95-617, 92 Stat. 3117 (1978) (codified at 16 U.S.C. § 2601 et seq.).

3. Cogeneration is a subset of distributed generation. The terms "cogeneration" and "combined heat and power" (CHP) are used interchangeably in this chapter. "Cogeneration" and "qualifying cogenerator" were in common use in Texas in the 1980s and 1990s. "CHP" is now the term used for reporting to the U.S. Department of Energy's Energy Information Administration. An EIA online glossary states, "CHP better describes the facilities because some of the plants included do not produce heat and power in a sequential fashion and, as a result, do not meet the legal definition of cogeneration specified in the Public Utility Regulatory Policies Act (PURPA)." http://www.eia.doe.gov/glossary/glossary_c.htm (accessed March 16, 2009).

4. An overview of section 210 of PURPA appears in Congressional Research Service, *Electricity: A New Regulatory Order?* Report prepared for the Committee on Energy and Commerce, U.S. House of Representatives, 102nd Cong., 1st sess., Committee Print 102-F (June 1991), 162–64.

5. Texas passed laws favorable to cogeneration in the sixty-seventh, sixty-eighth, sixty-ninth, and seventieth legislative sessions, 1981–87.

6. For technical and policy reasons, ERCOT and the PUCT restrict the definition of "distributed generation" to any generation resource directly connected to the electric distribution system (that is, below sixty kilovolts) that delivers less than ten megawatts to the ERCOT system. Units larger than ten MW are integrated into the ERCOT market. This chapter uses a broader definition of DG that does not depend on DG operation in synchronous or stand-alone mode, the interconnection voltage, power exports to ERCOT, the ownership restrictions set forth in PURPA, the type of technology or fuel source, the impact on the environment, or the annual hours or type of operation (for example, emergency, peaking, or baseload).

7. Tommy John, PE (vice president, Texas CHP Initiative), personal communications with the author, January 2008.

8. With the recent rapid growth of wind power in Texas, wind turbine farms—concentrations of generating units—have acquired the characteristics of central power plants, which require transmission investments to serve population centers. In Texas, wind turbine investments have recently necessitated public policies to address the recovery of major transmission investments.

9. In the eightieth session of the Texas legislature, HB 3693 introduced the concept of net metering in section 20. H.B. 3693, § 20, 80th Leg. Sess. (Tex. 2007),

codified as Tex. Util. Code § 39.107. Net metering is addressed in P.U.C.T. Substantive Rule § 25.242 16 Tex. Admin. Code § 25.242. The PUCT and ERCOT are clarifying the treatment of power from small-scale renewable generators as a result of HB 3693.

10. Twenty million vehicles in Texas multiplied by 25 kW per vehicle is equal to 500,000 MW, which is about five times the installed electric generation capacity in Texas. This comparison points to the wasted capability in the transportation sector (in that most vehicles sit idle most of the time), and the potential benefit to the electricity sector, if applying vehicle power to peak capacity needs becomes environmentally and financially reasonable. The municipal utility Austin Energy is experimenting with electric vehicle battery discharge as a peak capacity resource.

11. Companies such as EnerNOC and Comverge contract with customers to reduce usage by curtailing consumption or initiating backup generation upon request. Depending on the terms of the contract and the program established by the independent system operator (ISO), curtailing power may be driven by economics (the high cost of electricity) or by reliability (system emergencies; that is, capacity shortages due to loss of a large generating unit or transmission line) or both.

12. Texas Commission on Environmental Quality, "Air Quality Standard Permit for Electric Generating Units," May 2007.

13. Gulf Coast Regional CHP Application Center, *Proceedings of the Gulf Coast CHP Roadmap Workshop and Gulf Coast CHP Action Plan*, June 2005, 12–13.

14. See the discussion of the early electric industry in Texas in chapter 2 of Public Utility Commission of Texas, "The Scope of Competition in the Electric Industry in Texas: A Detailed Analysis," Report to the 75th Texas Legislature, January 1997, vol. II, II–7.

15. An "administered market" refers to a system of regulatory oversight that relies to a significant degree on the application of market forces. Replacing utility-constructed power plants with competitive procurement of contracts for power through all-source bidding is an example of an administered market.

16. Prior to the adoption of a formal methodology for calculating avoided capacity costs, the Public Utility Commission of Texas allowed utilities and large industrial customers to settle on a capacity payment of $3.00 per kilowatt month for firm power to qualifying cogenerators. Energy payments were in addition to this.

17. *Cogeneration and Small Power Monthly*, "Texas Issues Cogeneration Regulations," August 1, 1984.

18. *Cogeneration and Small Power Monthly*, "Texas Study Group on Cogeneration Issues Report on Avoided-Capacity Costs," August 1, 1984.

19. Unlike many physical commodities, electricity cannot be packaged and shipped from point A to point B. The flows on an alternating-current, high-voltage transmission line follow the laws of physics, not the terms of a legal contract. As with power exchanges between utilities, exchanges between nonutilities and utilities affect other parties on the transmission line. Wheeling agreements set forth the details and

compensation associated with generating more power at point A and taking more power out at point B, particularly for utilities that own transmission facilities between points A and B.

20. *Energy User News*, "Texas Lets Cogenerators Shop Around for Buyback," April 22, 1985.

21. See S.B. 141 and S.B. 142, 70th Leg. Sess. (Tex. 1987), reported in *Alternative Energy News*, "Texas Cogeneration Bills Become Law," July 1, 1987.

22. Pub. L. 102-486, 106 Stat. 2776 (1992).

23. Public Utility Regulatory Act of 1995, Tex. Rev. Civ. Stat. Ann. art. 1446c-0 §2.001(a); Senate Interim Committee on State Affairs, *Implementation of S.B. 373, Electric Utility Regulation*, available at http://www.senate.state.tx.us/SRC/75IntDig/senate/StAffair.htm (accessed March 16, 2009).

24. Adapted from Public Utilily Commission of Texas, "The Scope of Competition in the Electric Industry in Texas: A Detailed Analysis," Report to the 75th Texas Legislature, January 1997, vol. II, p. VI-34.

25. Ibid., VI-32, n.42.

26. P.U.C.T. Subst. R. § 25.223, 16 Tex. Admin. Code § 25.223 (1999).

27. P.U.C.T. Subst. R. § 25.341, 16 Tex. Admin. Code § 25.341 (2003).

28. P.U.C.T. Subst. R. § 25.343(c), 16 Tex. Admin. Code § 25.343(c) (2005).

29. As an employee of a regulatory agency, the author addressed instances where electric utilities expanded operations into competitive services. Competitive service providers were affected by utility actions that expanded service into activities that were not traditionally provided by utilities. These included the provision of lighting fixtures in commercial parking lots, the maintenance of residential cooling equipment, and the construction of electric substations for large industrial customers, to name a few specific cases.

30. See the open meeting decision adopting the cost-separation rule in *Rulemaking on Unbundling of Electric Distribution Facilities and Functions*, P.U.C.T. Project No. 16536 (August 26, 1998). Removing competitive services from regulation was subsequently addressed in rulemaking proceedings relating to the unbundling of energy service in 2000, available at http://www.puc.state.tx.us/rules/subrules/electric/25.342/25.342ei.cfm and http://www.puc.state.tx.us/rules/subrules/electric/25.343/25.343ei.cfm (accessed March 16, 2009).

31. Public Utility Regulatory Act of 1999 Tex. Rev. Civ. Stat. Ann. art. 1446c-0 Sec. 39.051.

32. State Energy Conservation Office (Texas), LoanSTAR Revolving Loan Program, online program description, http://www.seco.cpa.state.tx.us/seco_projects.htm (accessed February 11, 2009).

33. Amendments to Energy Efficiency Rules and Templates, new Substantive Rule § 25.181 relating to the Energy Efficiency Goal, P.U.C.T. Project No. 33487 (May 26, 2008).

34. P.U.C.T. Subst. R. §§ 25.211 and 25.212 16 Tex. Admin. Code §§ 25.211 and 25.212.

35. P.U.C.T. Subst. R. § 25.211(n), 16 Tex. Admin. Code § 25.211(n).

36. Standardization of Small Generator Interconnection Agreements and Procedures, Order No. 2006, 70 Fed. Reg. 34100 (Jun. 13, 2005), F.E.R.C. Stats. & Regs., Regulations Preambles, Vol. III, ¶ 31,180, ¶¶ 31,406-31,551 (2005).

37. Tex. Util. Code § 39.916 (2007), added as part of H.B. 3963.

38. *Net Metering and Interconnection of Distributed Generation*, P.U.C.T. Project No. 34890.

39. P.U.C.T. Subst. R. § 25.242, 16 Tex. Admin. Code § 25.242.

40. David Dismukes, "The Role of Distributed Energy Resources in a Restructured Power Industry," in *Electric Choices: Deregulation and the Future of Electric Power*, ed. Andrew Kleit (Oakland, CA: Rowman and Littlefield, 2006), 197.

41. Nat Treadway, "Rethinking Regulated Ratemaking: Menu-of-Service Electric Utility Tariffs," in *Proceedings of the 2005–2006 IEEE/PES Transmission and Distribution Conference and Exposition*, Dallas, May 21–24, 2006.

42. An example of menu-of-service pricing is load-limiting service, which is common in Europe and has been successfully deployed in Texas. Customers pay less but receive a maximum level of power that requires load-management strategies. The issues related to menu-of-service pricing are presented in Center for the Advancement of Energy Markets, *Analysis and Development of Model Options of Electric Utility Rates and Tariffs Affecting DER, Task 1: Statement of the Problem and Issues*, National Renewable Energy Laboratory, Subcontract Number AAT-4-32616-01, May 2005. The author was the primary researcher for that report.

43. The Texas State Implementation Plan filed with the U.S. Environmental Protection Agency has undergone numerous changes and updates since 1972. The plans and revisions are available on the TCEQ website: http://www.tceq.state.tx.us/implementation/air/sip/sipplans.html (accessed March 16, 2009).

44. Clean Air Act, 42 U.S.C. 7401 et seq.

45. Texas Commission on Environmental Quality, "Air Quality Standard Permit for Electric Generating Units," May 2007, www.tceq.state.tx.us/assets/public/permitting/air/NewSourceReview/Combustion/egu_final_spermit.pdf (accessed March 16, 2009).

46. Ibid. To qualify for a standard permit in the East Texas Region, a generating unit with a capacity of 250 kW or more must emit less than 0.14 pounds of nitrogen oxides per megawatt-hour of electricity output (0.14 lb/MWh; the East Texas Region is defined as counties traversed by or east of Interstate Highway 35 or Interstate Highway 37, including Bosque, Coryell, Hood, Parker, Somervell, and Wise Counties). This rate is approximately equivalent to five parts per million by volume of nitrogen oxides (5 ppmv NOx), a more common measure in other jurisdictions. Smaller units, with a capacity of 250 kW or less, must meet a standard of 0.47 lb/MWh of NOx to use the standard permit for East Texas Region DG installations. Electric generating units that operate for fewer than 300 hours per year and those that use waste products for fuel are subject to less stringent standards. There are also provisions to facilitate the permitting of CHP units. A CHP unit that recovers heat comprising more than 20 percent of its total output and that satisfies certain

documentation requirements may take credit for the recovered heat. The standard permit states, "Credit shall be at the rate of one MWh for each 3.4 million British Thermal Units of heat recovered."

47. The energy-efficiency resource standard would have included CHP and other types of energy efficiency. See the Texas Combined Heat and Power Initiative, "Twice the Power at Double the Efficiency: Providing Secure Energy in Texas with CHP," February 2007, or the Texas CHP Initiative legislative action sheet, available at http://www.texaschpi.org (accessed March 16, 2009).

48. P.U.C.T. Subst. R. § 25.173(g), 16 Tex. Admin. Code § 25.173(g).

Chapter 7: Competitive Performance of the ERCOT Wholesale Market

1. For detailed descriptions and statistics of the wholesale market, see Ross Baldick and Hui Niu, "Lessons Learned: The Texas Experience," in *Electricity Deregulation: Where to from Here?* ed. James Griffin and Steven Puller (Chicago: University of Chicago Press, 2005), and annual ERCOT state-of-the-market reports by Potomac Economics; see http://www.potomaceconomics.com/documents/C6&C10, and citations for specific reports, below.

2. Transmission congestion has been priced on a zonal basis, but ERCOT plans to move to nodal pricing in 2009; see chapter 3.

3. Potomac Economics, *2006 State of the Market Report for the ERCOT Wholesale Electricity Markets*, August 2007: see http://www.potomaceconomics.com/uploads/ercot_reports/2006%20ERCOT%20SOM%20REPORT_Final.pdf (accessed March 24, 2009).

4. The theory and evidence of the "energy-only" approach are detailed in chapter 4, above.

5. This measure of marginal cost would include the variable cost of fuel, operating and maintenance, and emissions permits. Marginal cost also includes the opportunity cost of selling power to other markets (or, in the case of reservoir generation, in different time periods).

6. See Severin Borenstein, James B. Bushnell, and Frank A. Wolak, "Measuring Market Inefficiencies in California's Restructured Wholesale Electricity Market," *American Economic Review* 92, no. 5 (2002): 1376–1405, and Erin T. Mansur, "Measuring Welfare in Restructured Electricity Markets," *Review of Economics and Statistics* 90, no. 2 (May 2008): 369–86 for examples of comparing actual prices to the perfectly competitive benchmark.

7. Under a pool model, generators and load-serving entities may also sign contracts to hedge against the pool price.

8. For a discussion of the theory behind transmission rights, see William W. Hogan, "Market-Based Transmission Investments and Competitive Electricity Markets," in *Electric Choices: Deregulation and the Future of Electric Power*, ed. Andrew N. Kleit (Oakland, CA: Rowman and Littlefield, 2007).

9. Potomac Economics, *2006 State of the Market Report.*

10. For further discussion of market monitoring in ERCOT, see chapter 9.

11. For example, see Borenstein, Bushnell, and Wolak, "Measuring Market Inefficiencies," and Mansur, "Measuring Welfare in Restructured Electricity Markets."

12. And, for simplicity, assume that there is no local or interzonal congestion and that ramping constraints do not bind.

13. Ali Hortaçsu and Steven L. Puller, "Understanding Strategic Bidding in Multi-Unit Auctions: A Case Study of the Texas Electricity Spot Market," *RAND Journal of Economics* 39, no. 1 (2008): 86–114.

14. This analysis may appear to assume that firms know their rivals' bids before submitting their own. Bidders do not know current rivals' bids, but they do have access to rivals' bids with a two- to three-day lag. See Hortaçsu and Puller, "Understanding Strategic Bidding," for evidence that the use of publicly available bid data can be used to achieve profits very close to those of the method described here.

15. Specifically, we focused on the interval 18:00–18:15 in the 75 percent of the intervals with no interzonal transmission congestion.

16. See Public Utility Regulatory Act, Tex. Util. Code § 39.1515 (2006) and P.U.C.T. Subst. R. § 25.504, 16 Tex. Admin. Code § 25.504.

17. Ramteen Sioshansi and Shmuel Oren, "How Good Are Supply Function Equilibrium Models? An Empirical Analysis of the ERCOT Balancing Market," *Journal of Regulatory Economics* 31, no. 1 (February 2007): 1–35.

Chapter 8: Retail Restructuring and Market Design in Texas

1. Senate Bill 7, "An Act Relating to Utility Restructuring and to the Powers and Duties of the Public Utility Commission of Texas, Office of Public Utility Counsel, and Texas Natural Resource Conservation Commission; providing penalties," 1999. Available online at http://www.capitol.state.tx.us/tlodocs/76R/billtext/htim/SB00007F.htm (accessed March 31, 2009).

2. Available online at http://tlo2.tlc.state.tx.us/statutes/ut.toc.htm (accessed March 31, 2009).

3. Public Utility Commission of Texas, "Electric Power Industry Scope of Competition and Potential Stranded Investment Report, Volume I," Report to the 75th Texas Legislature (1997), 12 . Many of these facilities are refiners and petrochemical facilities concentrated in Port Arthur, the Houston Ship Channel, and Corpus Christi.

4. Ibid., 17–18.

5. Note that, as in the market design in the privatization of the electricity industry in England and Wales in 1990, this design created a stand-alone wires company that would continue to be regulated by the PUCT.

6. Texas Senate Bill 7, op. cit., Sec. 39.202.

7. Small commercial customers were defined as those with peak demand of less than 1 MW.

8. The number of incumbent territories was five or six, depending on the treatment of two separate TXU affiliates. This analysis uses two different classifications: five AREP territories (TXU ED, Centerpoint/Reliant, AEP Central, AEP North, and TNMP), or six AREP territories (TXU Energy, Reliant Energy, First Choice Power, West Texas Utilities, Central Power and Light, and TXU Sesco).

9. California's standard-offer price was also the result of an administrative cost-allocation procedure that did not incorporate any headroom, new entrant profit margin, or return to the retail sales function. Lynne Kiesling, *Getting Electricity Deregulation Right: How Other States and Nations Have Avoided California's Mistakes,* Comment to the Federal Trade Commission on Retail Electricity Competition Plans (2001), available at www.ftc.gov/os/comments/eleccompetition/reasoninst.pdf; see also the *Retail Electricity Deregulation (RED) Index 2000, Center for the Advancement of Energy Markets.*

10. Public Utility Commission of Texas, "Report on the Scope of Competition in Electricity Markets in Texas," Report to the 78th Texas Legislature (2003), 23.

11. In 2003, the PUCT revised the threshold to a 5 percent change in the twenty-day average of the forward twelve-month NYMEX Henry Hub natural gas contract.

12. Public Utility Commission of Texas, "Report on the Scope of Competition" (2003), 29.

13. Texas Senate Bill 7, op. cit., Sec. 39.904.

14. Ibid., 59.

15. See Public Utility Commission of Texas, "Texas Electric Choice Education Program," www.powertochoose.org and www.poderdeescoger.org (accessed February 12, 2009).

16. In 2004 and 2005, the Texas legislature reduced SBF funding and did not appropriate any funds for weatherization or property tax replacement. Public Utility Commission of Texas, "Report on the Scope of Competition in Electricity Markets in Texas," Report to the 79th Texas Legislature (2005), 41.

17. Public Utility Commission of Texas, "Report on the Scope of Competition" (2003), 79–80.

18. Public Utility Commission of Texas, "Report on the Scope of Competition" (2005), 44.

19. Public Utility Commission of Texas, "Report on the Scope of Competition in Electricity Markets in Texas," Report to the 80th Texas Legislature (2007), 8.

20. Ibid., 50.

21. Note also that SB 7 included a demand-reduction goal of 135 MW to increase energy efficiency; this goal was achieved by 2004, with 151 MW of demand reduction attributable specifically to energy-efficiency initiatives. Moreover, most of this demand reduction occurred in areas that were EPA-designated nonattainment or near-nonattainment areas for air quality. Public Utility Commission of Texas, "Report on the Scope of Competition" (2005), 68.

22. Public Utility Commission of Texas, "Report on the Scope of Competition" (2003), 100.

23. Public Utility Commission of Texas, "Report on the Scope of Competition" (2007), 18. See also chapter 5, above.

24. Public Utility Commission of Texas, "Report on the Scope of Competition" (2003), 96–98.

25. Ibid., 62.

26. Bob Richter, "Deregulation of Utilities Criticized," *San Antonio Express-News*, May 28, 2002, 1B; David C. Johnston, "Electric Rates Haven't Fallen," *Fort Worth Star-Telegram*, October 15, 2006, A6; R. A. Dyer, "Voters Worried about Rates," *Fort Worth Star-Telegram*, October 23, 2006, B1; Dan Piller, "Higher Rates, Profit for TXU," *Fort Worth Star-Telegram*, February 3, 2006, C1; *Corpus Christi Caller-Times*, "Lawsuit Alleges CPL Rates Excessive," April 23, 2002, D7; Sudeep Reddy, "Tangled Up in Higher Prices but Supporters Say Texans Will Benefit in Time," *Dallas Morning News*, December 30, 2004, 1D; Tom Fowler, "Higher Power Bills through Competition Fuel Rules Mean Many Aren't Seeing the Lower Rates They Expected," *Houston Chronicle*, June 12, 2005, 1.

27. Harold Demsetz, "Why Regulate Utilities?" *Journal of Law and Economics* 11 (April 1968): 55–65.

28. Public Utility Commission of Texas, "Report on the Scope of Competition" (2007).

29. Electricity Reliability Council of Texas, *2006 Annual Report*, available at http://www.ercot.com/news/presentations/2007/2006_Annual_Report.pdf (accessed February 12, 2009).

30. Julie Parsley, "What Have You Done for Me Lately? A Look at the Texas Competitive Electric Market," briefing to the Texas legislature, November 29, 2007, http://www.puc.state.tx.us/about/commissioners/parsley/present/epp/Comp_Elec_Market_112907.pdf (accessed February 12, 2009).

31. Moreover, under regulation the fuel cost changes flow through to retail rates only with a substantial lag, after a possibly lengthy rate case. In markets, the quick reflection of cost increases in retail prices is one of the hallmarks of efficiency.

32. Public Utility Commission of Texas, "Report on the Scope of Competition in Electricity Markets in Texas," Report to the 79th Texas Legislature (2006), 2.

Chapter 9: Market Monitoring, ERCOT Style

1. Harold Demsetz, "The Trust behind Antitrust" in *Efficiency, Competition, and Policy*, vol. 2 of *The Organization of Economic Activity*, ed. Harold Demsetz, 217–24 (Oxford: Blackwell, 1989).

2. See, for example, Severin Borenstein, James B. Bushnell, and Frank A. Wolak, "Measuring Market Inefficiencies in California's Restructured Wholesale Electricity Market," *American Economic Review* 92, no. 5 (December 2002): 1376–1405; Steven L. Puller, "Pricing and Firm Conduct in California's Deregulated Electricity Market," *Review of Economics and Statistics* 89, no. 1 (February 2007): 75–87; and Andrew

Sweeting, "Market Power in the England and Wales Wholesale Electricity Market 1995–2000," *Economics Journal* 117, no. 520 (April 2007): 654–85.

3. See, for example, L. Lynne Kiesling, "The Role of Retail Pricing in Electricity Restructuring," in *Electric Choices: Deregulation and the Future of Electric Power*, ed. Andrew N. Kleit, 39–62 (New York: Rowman and Littlefield, 2007).

4. This lack of demand elasticity may recede over time as new monitoring devices are installed.

5. Ideally, one would want to use marginal costs to show the "hockey stick." Unfortunately, I have been unable to obtain a marginal cost curve of production for ERCOT. The displayed stack of bids should be fairly representative, however.

6. See, for example Andrew Kleit, Blake Nelson and John Rohrback, "Can FERC Solve Its Market Power Problems? Supply Margin Assessment Doesn't Seem to Be a Promising First Step," *The Electricity Journal* 15:3 (2002): 10–18; and Darren Bush and Carrie Mayne, *In Reluctant Defense of Enron: Why Bad Regulation Is to Blame for California's Power Woes (or Why Antitrust Law Fails to Protect against Market Power When the Market Rules Encourage Its Use)*, 83 Or. L. Rev. 207, 277–80 (2004).

7. See, for example, Severin Borenstein, James Bushnell, and Christopher R. Knittel, "Market Power in Electricity Markets: Beyond Concentration Measures," *Energy Journal* 20, no. 4 (1999): 65–88.

8. See, for example, the discussion in Bush and Mayne, "In Reluctant Defense of Enron."

9. See, for example, Harold Demsetz, "Information and Efficiency: Another Viewpoint," *Journal of Law and Economics* 12, no. 1 (1969): 1–22.

10. *Texas Comm. Energy v. TXU*, 413 F.3d 503 (5th Cir. 2005). See chapter 1, above.

11. James D. Reitzes, Johannes P. Pfeifenberger, Peter Fox-Penner, Gregory N. Basheda, José A. García, Samuel A. Newell, and Adam C. Schumacher, "Review of PJM's Market Power Mitigation Practices in Comparison to Other Organized Electricity Markets," September 14, 2007, www.brattle.org/_documents/UploadLibrary/Upload631.pdf (accessed March 9, 2009).

12. Michael Dworkin and Rachel Goldwasser, "Ensuring Consideration of the Public Interest in the Governance and Accountability of Regional Transmission Organizations," *Energy Law Journal* 28 (2007): 543.

13. John McNeely Foster and William R. Mayben, Report of Investigation by the Special Investigative Committee of the Board of Managers of PJM Interconnection, L.L.C, Frank L. Olson, Chair, November 7, 2007, 6, www2.pjm.com/documents/mmu/pjm-special-investigative-committee-report-20071107.pdf (accessed March 9, 2009).

14. PJM and various customer groups reached a settlement on the market monitoring issue, which was submitted to FERC on December 19, 2007. See www.brattle.org/_documents/UploadLibrary/Upload631.pdf (accessed March 6, 2009). Dr. Bowring would direct market monitoring as the principal of an outside firm with a contract with PJM, under which PJM would agree to supply Bowring with certain market data.

15. The California and New York ISOs are the other state-specific RTOs currently in operation.

16. Parviz Adib (director of the original PUCT staff market-monitoring unit), conversation with the author, November 12, 2007. Discomfort with the ISO-NE example also appears to be a reason for there currently being no bid mitigation in ERCOT.

17. Dan Jones, in an interview by Andrew Kleit, November 13, 2007, Austin, Texas.

18. See Reitzes et al., "Review of PJM's Market Power Mitigation Practices," and L. Lynne Kiesling and Bart J. Wilson, "An Experimental Analysis of the Effects of Automated Mitigation Procedures on Investment and Prices in Wholesale Electricity Markets," *Journal of Regulatory Economics* 31, no. 3 (2007): 313–34.

19. Timothy J. Brennan, "Preventing Monopoly or Discouraging Competition: The Perils of Price Cost Tests in Electricity," in *Electric Choices*, 169.

20. See, for example, the discussion in P.U.C.T. Substantive Rule § 25.504, http://www.puc.state.tx.us/rules/subrules/electric/index.cfm.

21. See David Hurlbut, Keith Rogas, and Shmuel Oren, "Protecting the Market from Hockey Stick Pricing: How the Public Utility Commission of Texas Is Dealing with Potential Price Gouging," *Electricity Journal* 17, no. 3 (April 2004): 26–32.

22. Ali Hortaçsu and Steven L. Puller, "Understanding Strategic Bidding"; Ramteen Sioshansi and Shmuel Oren, "How Good Are Supply Function Equilibrium Models: An Empirical Analysis of the ERCOT Balancing Market," *Journal of Regulatory Economics* 31, no. 1 (February 2007): 1–35.

23. This conclusion is somewhat discouraging, as it implies that, over time, firms will not learn what is in their best interest, and a regulator will need to intervene to correct the market failure.

24. Public Utility Commission of Texas, *Staff Inquiry into Allegations Made by Texas Commercial Energy regarding ERCOT Market Manipulation*, P.U.C.T. Project No. 25937 (January 28, 2004), www.puc.state.tx.us/WMO/documents/special/TCE_allegations.pdf (accessed March 6, 2009).

25. Looking, for example, at the bid stack for January 31, 2007, hour 9 (available from ERCOT), the 95 percent bid is $187, which would imply a "mitigated" price of 1.5*$187=$280.5/MWh.

26. It should be noted that publicizing the identity of "bad actors" has worked well to reduce emissions in the environmental context. See, for example, Shameek Konar and Mark A. Cohen, "Information as Regulation: The Effect of Community Right to Know Laws on Toxic Emissions," *Journal of Environmental Economics and Management* 32, no. 1 (January 1997): 109–24.

27. *Staff Inquiry into Allegations Made by Texas Commercial Energy regarding ERCOT Market Manipulation*, P.U.C.T. Project No. 25937 (January 28, 2004), 5.

28. Ibid., 6.

29. See *Rulemaking Concerning Resource Adequacy and Market Power in the Electric Reliability Council of Texas Power Region*, P.U.C.T. Rulemaking No. 31972 (October 2006), available at http://www.puc.state.tx.us/rules/rulemake/31972/31972.cfm.

30. Brennan notes that under a regime of marginal-cost pricing, the producer with the highest marginal costs (the "peaking" plant) cannot cover its fixed costs. This critique, however, does not appear to apply to the TXU matter, which does not involve peaking units. Brennan, "Preventing Monopoly," 166.

31. Potomac Economics, "Investigation of the Wholesale Market Activities of TXU from June 1 to September 30, 2005," March 2007, 22.

32. Initial Staff Complaint, *Notice of Violation by TXU*, P.U.C.T. Control File No. 34061, #1, at 2 (March 28, 2007).

33. Revised Staff Complaint, *Notice of Violation by TXU*, P.U.C.T. Control File No. 34061, #105, at 5 (September 14, 2007).

34. See Response of TXU, *Notice of Violation by TXU*, P.U.C.T. Control File No. 34061, #167 (November 9, 2007). More generally, see Bush and Mayne, "In Reluctant Defense of Enron."

35. See Response of TXU, P.U.C.T. Control File No. 34061, at 14.

36. PUCT Subst. R. § 25.504(b)(3), 16 Tex. Admin. Code § 25.504(b)(3).

37. PUCT Subst. R. § 25.504(d), 16 Tex. Admin. Code § 25.504(d).

38. *TXU v. Public Util. Comm. of Texas*, 165 S.W.3d 821 (Tex. App. 2005).

39. Ibid., 836–37.

40. Ibid., 829.

41. TXU Wholesale Companies' Request for Approval of a Voluntary Mitigation Plan, Control File 34480, #19, July 27, 2007, http://interchange.puc.state.tx.us/WebApp/Interchange/application/dbapps/filings/pgControl.asp?TXT_UTILITY_TYPE=A&TXT_CNTRL_NO=34480&TXT_ITEM_MATCH=1&TXT_ITEM_NO=&TXT_N_UTILITY=&TXT_N_FILE_PARTY=&TXT_DOC_TYPE=ALL&TXT_D_FROM=&TXT_D_TO=&TXT_NEW=true (accessed March 16, 2009).

42. Unfortunately, because this was not a contested proceeding, the PUCT did not offer any rationale for the settlement.

43. See chapter 7.

44. One commenter on this chapter suggested that the VMP would have been valuable because TXU would have been required to offer its units at some price, eliminating the possibility of physical withholding. As discussed above, however, the impact of physical holding can be replicated by simply making above-market offers.

45. See *Reliant. v. Public Util. Comm. of Texas*, No.D-1-GN-07-002494 (345th District Court of Texas, October 7, 2007).

Index

About the Authors

Parviz Adib is a director in APX's Professional Services group. He has worked with energy markets for more than twenty-five years. Prior to joining APX, he worked at the Public Utility Commission of Texas for twenty-one years and served as the market monitor for the Texas electricity market. Earlier, he worked in the Bureau of Business Research at the University of Texas and served in supervisory positions for both Crest Engineering and International Harvester. He holds a Ph.D. in economics from the University of Texas at Austin.

Darren Bush is associate professor of law at the University of Houston Law Center where he focuses on antitrust and deregulated industries. Professor Bush previously served as Attorney General's Honor Program Trial Attorney at the Antitrust Division's Transportation, Energy, and Agriculture Section. He has a J.D. and a Ph.D. in economics from the University of Utah.

Gürcan Gülen is senior energy economist, Center for Energy Economics, Bureau of Economic Geology, the University of Texas at Austin. He previously worked at the University of Houston as an energy economist. He received a Ph.D. in economics from Boston College and a B.A. in economics from Bosphorus University in Istanbul, Turkey.

L. Lynne Kiesling is senior lecturer in the Department of Economics at Northwestern University and in the Social Enterprise at Kellogg (SEEK) program in the Kellogg School of Management at Northwestern University. She is also currently a senior research engineer at the Pacific Northwest National Laboratory and a member of the GridWise Architecture Council. She was previously a research scholar at the Interdisciplinary Center for Economic Science at George Mason University, director of economic policy at the Reason

Foundation, manager at PricewaterhouseCoopers LLP, and assistant professor at the College of William and Mary. She has a Ph.D. in economics from Northwestern University.

Andrew N. Kleit is professor of energy and environmental economics at Pennsylvania State University. He was previously associate professor of economics at Louisiana State University, senior economic adviser to the director for Investigation and Research (the chief antitrust official in Canada), and economic adviser to the director, Bureau of Competition, Federal Trade Commission. He has a Ph.D. from Yale University.

Shmuel S. Oren is the Earl J. Isaac Professor in the Science and Analysis of Decision Making in the Department of Industrial Engineering and Operations Research at the University of California at Berkeley and former chairman of that department. He is the Berkeley site director of the Power System Engineering Research Center (PSerc). He has been an adviser to staff of the Public Utility Commission of Texas (PUCT) since 2001 and is currently a consultant to the energy division of the California Public Utility Commission. He holds a Ph.D. from Stanford in engineering economic systems.

Brett A. Perlman served as commissioner of the Public Utility Commission of Texas from 1999 to 2003 and was the only commissioner to serve at PUCT from the passage of the Texas electric market restructuring legislation through the opening of the Texas wholesale and retail electricity market. He is currently president of Vector Advisors, a management consulting firm that provides services to telecommunications and energy clients. He holds a law degree from the University of Texas and a master's degree in public administration from the John F. Kennedy School of Government at Harvard University.

Steven L. Puller is associate professor of economics at Texas A&M University and a faculty research fellow with the National Bureau of Economic Research. He also has served as a visiting research associate at the University of California Energy Institute and an adviser to the Public Utility Commission of Texas. He has a Ph.D. in economics from the University of California at Berkeley.

Eric S. Schubert is regulatory affairs adviser for BP Energy Company. His professional experience includes working at the Public Utility Commission of Texas, the Chicago Board of Trade, and Bankers Trust Company. During his tenure at the PUCT, Eric was project leader in proceedings involving wholesale market design, resource adequacy, and renewables issues in ERCOT. He has a Ph.D. in economics from the University of Illinois at Urbana-Champaign.

David Spence is associate professor of law, politics, and regulation at the University of Texas at Austin's McCombs School of Business, where he teaches courses on energy regulation, environmental regulation, and business-government relations. He holds advanced degrees from Duke University (Ph.D.) and the University of North Carolina (J.D.), and has taught at Vanderbilt University Law School, the Harvard Law School, the Cornell University School of Law, the Bren School of Environmental Management at the University of California at Santa Barbara, the Nicholas School of the Environment at Duke University, Edinburgh University (Scotland), and IMADEC University (Vienna, Austria).

Jess Totten is the director of the Competitive Markets Division for the Public Utility Commission of Texas. Since 1999, he has managed the implementation of the Texas retail competition law and related programs, such as renewable energy and energy efficiency. He has also worked to develop legislation and policy relating to retail and wholesale competition in the electric industry, electric transmission, energy efficiency, and renewable energy. Prior to joining the Texas PUC, he served as an attorney and deputy general counsel for the Panama Canal Commission. Mr. Totten received a law degree from the University of Texas.

Nat Treadway is a managing partner of the Distributed Energy Financial Group, LLC (www.defgllc.com), which provides management consulting, strategic marketing (through EcoAlign, www.ecoalign.com), and venture capital (DEFG Ventures, LLC). He is a member of the board of directors of the Texas Combined Heat and Power Initiative. He was previously a senior policy adviser at the Public Utility Commission of Texas. He holds a master's degree in agricultural economics from Michigan State University.

Pat Wood III is principal of Wood3 Resources, an energy development firm in Houston. He is past chairman of the Federal Energy Regulatory Commission and of the Public Utility Commission of Texas, where he promoted well-organized markets and infrastructure investment. Actively developing clean power projects, he also serves as a director of a number of public and private companies in the energy sector. He holds a law degree from Harvard University.

The American Enterprise Institute for Public Policy Research

Founded in 1943, AEI is a nonpartisan, nonprofit research and educational organization based in Washington, D.C. The Institute sponsors research, conducts seminars and conferences, and publishes books and periodicals.

AEI's research is carried out under three major programs: Economic Policy Studies, Foreign Policy and Defense Studies, and Social and Political Studies. The resident scholars and fellows listed in these pages are part of a network that also includes ninety adjunct scholars at leading universities throughout the United States and in several foreign countries.

The views expressed in AEI publications are those of the authors and do not necessarily reflect the views of the staff, advisory panels, officers, or trustees.

www.ingramcontent.com/pod-product-compliance
Lightning Source LLC
Jackson TN
JSHW011933131224
75386JS00041B/1360
* 9 7 8 0 8 4 4 7 4 2 8 2 3 *